Predictable and Runtime-Adaptable Network-On-Chip for Mixed-critical Real-time Systems

Dissertation an der Technischen Universität Braunschweig,
Fakultät für Elektrotechnik, Informationstechnik, Physik

I0054504

Predictable and Runtime-Adaptable Network-On-Chip for Mixed-critical Real-time Systems

Von der Fakultät für Elektrotechnik, Informationstechnik, Physik
der Technischen Universität Carolo-Wilhelmina zu Braunschweig

zur Erlangung des Grades eines Doktors

der Ingenieurwissenschaften (Dr.-Ing.)

genehmigte Dissertation

von Sebastian Tobuschat

aus Pinneberg

Eingereicht am: 12.12.2018

Mündliche Prüfung am: 07.02.2019

1. Referent: Prof. Dr.-Ing. Rolf Ernst
2. Referent: Prof. Dr.-Ing. Dr. h. c. Jürgen Becker

Druckjahr: 2019

Bibliographische Information der Deutschen Nationalbibliothek
Die Deutsche Nationalbibliothek verzeichnet diese Publikation in der
Deutschen Nationalbibliographie; detaillierte bibliographische Daten
sind im Internet über http://dnb.d-nb.de abrufbar.
1. Aufl. - Göttingen: Cuvillier, 2019
   Zugl.: (TU) Braunschweig, Univ., Diss., 2019

© CUVILLIER VERLAG, Göttingen 2019
   Nonnenstieg 8, 37075 Göttingen
   Telefon: 0551-54724-0
   Telefax: 0551-54724-21
   www.cuvillier.de

ISBN 978-3-7369-9979-4
eISBN 978-3-7369-8979-5

Abstract

The industry of safety-critical and dependable embedded systems calls for even cheaper, high performance platforms that allow flexibility and an efficient verification of safety and real-time requirements. In this sense, flexibility denotes the ability to (online) adapt a system to changes (e.g. changing environment, application dynamics, errors) and the reuse-ability for different use cases. To cope with the increasing complexity of interconnected functions and to reduce the cost and power consumption of the system, multicore systems are used to efficiently integrate different processing units in the same chip. Networks-on-chip (NoCs), as a modular interconnect, are used as a promising solution for such multiprocessor systems on chip (MPSoCs), due to their scalability and performance. Hence, future NoC designs must face the aforementioned challenges.

For safety-critical systems, a major goal is the avoidance of hazards. For this, safety-critical systems are qualified or even certified to prove the correctness of the functioning under all possible cases. A predictable behaviour of the NoC can help to ease the qualification process (e.g. formal analysis) of the system. To achieve the required predictability, designers have two classes of solutions: isolation (quality of service (QoS) mechanisms) and (formal) analysis. For mixed-criticality systems, isolation and analysis approaches must be combined to efficiently achieve the desired predictability. Isolation techniques are used to bound interference between different application classes. And analysis can then be applied verifying the real-time applications and *sufficient isolation* properties.

Traditional NoC analysis and architecture concepts tackle only a subpart of the challenges—they focus on either performance or predictability. Existing, predictable NoCs are deemed too expensive and inflexible to host a variety of applications with opposing constraints. And state-of-the-art analyses neglect certain platform properties (e.g. they assume sufficient buffer sizes to avoid backpressure) to verify the behaviour. Together this leads to a high over-provisioning of the hardware resources as well as adverse impacts on system performance (especially for the non safety-critical applications), and on the flexibility of the system.

In this work we tackle these challenges and develop a predictable and runtime-adaptable NoC architecture that efficiently integrates mixed-critical applications with opposing constraints. Additionally, we present a modelling and analysis framework for NoCs that accounts for backpressure (i.e. full buffers in network routers delaying the progress of network packets). This framework enables to evaluate the performance and reliability early at design time. Hence, the designer can assess multiple design decisions and trade-offs (such as area, voltage, reliability, performance) by using abstract models and formal approaches.

Zusammenfassung

Die Industrie der sicherheitskritischen und zuverlässigen eingebetteten Systeme verlangt nach noch günstigeren, leistungsfähigeren Plattformen, welche Flexibilität und eine effiziente Überprüfung der Sicherheits- und Echtzeitanforderungen ermöglichen. Flexibilität bezeichnet in diesem Sinne die Fähigkeit, ein System zur Laufzeit an Veränderungen (z.b. sich verändernde Umgebungen, Anwendungsdynamik, Fehler) anzupassen als auch die Wiederverwendbarkeit für verschiedene Anwendungsfälle. Um der zunehmenden Komplexität der zunehmend vernetzten Funktionen gerecht zu werden und die Kosten und den Stromverbrauch eines Systems zu reduzieren, werden Mehrkern-Systeme eingesetzt. On-Chip Netzwerke (NoCs) werden aufgrund ihrer Skalierbarkeit und Leistung als vielversprechende Lösung für solch Mehrkern-Systeme eingesetzt. Daher müssen sich zukünftige on-Chip Netzwerke den oben genannten Herausforderungen stellen.

Bei sicherheitskritischen Systemen ist die Vermeidung von Gefahren ein wesentliches Ziel. Dazu werden sicherheitskritische Systeme qualifiziert oder zertifiziert, um die Funktionsfähigkeit in allen möglichen Fällen nachzuweisen. Ein vorhersehbares Verhalten des on-Chip Netzwerks kann dabei helfen, den Qualifizierungsprozess (z.B. die formale Analyse) des Systems zu erleichtern. Um die erforderliche Vorhersagbarkeit zu erreichen, gibt es zwei Klassen von Lösungen: Isolation (Quality of Service Mechanismen) und (formale) Analyse. Für Systeme mit gemischter Relevanz müssen Isolationsmechanismen und Analyseansätze kombiniert werden, um die gewünschte Vorhersagbarkeit effizient zu erreichen. Isolationsmechanismen werden eingesetzt, um Interferenzen zwischen verschiedenen Anwendungsklassen zu begrenzen. Und die Analyse wird angewendet, um die Echtzeitfähigkeit und eine *hinreichende Isolation* zu verifizieren.

Traditionelle Analyse- und Architekturkonzepte für on-Chip Netzwerke lösen nur einen Teil dieser Herausforderungen—sie konzentrieren sich entweder auf Leistung oder Vorhersagbarkeit. Existierende vorhersagbare on-Chip Netzwerke werden als zu teuer und unflexibel erachtet, um eine Vielzahl von Anwendungen mit gegensätzlichen Anforderungen zu integrieren. Und state-of-the-art Analysen vernachlässigen bzw. vereinfachen bestimmte Plattformeigenschaften (beispielsweise wird eine ausreichend große Puffergröße angenommen, um einen Paketrückstau zu vermeiden), um das Verhalten überprüfen zu können. Dies führt zu einer hohen Überbereitstellung der Hardware-Ressourcen als auch zu negativen Auswirkungen auf die System-leistung (insbesondere für die nicht sicherheitskritischen Anwendungen) und auf die Flexibilität des Systems.

In dieser Arbeit gehen wir auf diese Herausforderungen ein und entwickeln eine vorhersehbare und zur Laufzeit anpassbare Architektur für on-Chip Netzwerke, welche gemischt-kritische Anwendungen effizient integriert. Zusätzlich stellen wir ein Modellierungs- und Analyseframework für on-Chip Netzwerke vor, das den Paketrückstau berücksichtigt (d.h. die Puffer im Netzwerk dürfen überlaufen). Dieses Framework ermöglicht es, die Leistung und Zuverlässigkeit bereits zur Designzeit zu bewerten. Somit kann der Konstrukteur mehrere Designentscheidungen und Kompromisse (wie Fläche, Spannung, Zuverlässigkeit, Leistung) anhand abstrakter Modelle und formaler Ansätze frühzeitig beurteilen.

To Marlen.

Contents

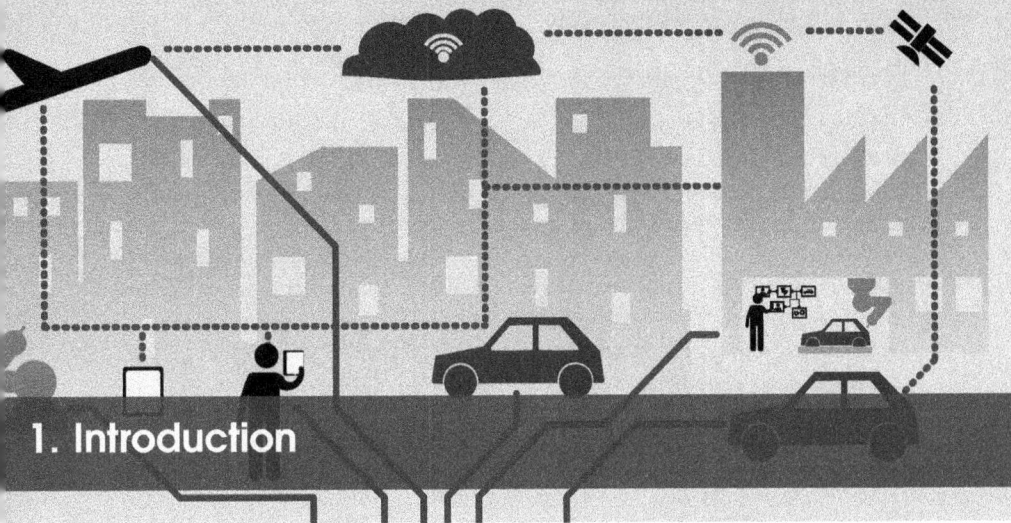

1. Introduction

1.1 Motivation

Safety-critical and dependable embedded systems play an important role in our daily life. For example, modern vehicles integrate over 100 electronic control units (ECUs). Due to the ever-increasing demand for high performance together with low energy consumption and size, multicore systems, as known from general-purpose computing, are adopted by the safety-critical embedded market. The integration of multiple cores in a chip to a multiprocessor system on chip (MPSoC) offers the possibility to consolidate multiple functions or ECUs, which previously had been distributed and isolated by external buses. This consolidation of functions with different overall importance leads to mixed-criticality multicore systems [28]. In this context, a critical function is essential for the safety of the system. Therefore, this function is developed with high diligence and so the behaviour (e.g. timing) is well specified and tested. For non-critical functions the confidence in the characteristics is lower, e.g., the possibility that the function deviates from the specification is higher. Additionally, non-critical functions might be user provided and the risk of malicious functions trying to endanger system safety, e.g., through denial-of-service attacks, increases.

Figure 1.1 presents an example for typical features in a modern car. These include classical applications, such as engine control or entertainment functions, but also new complex functions for highly automated and autonomous driving, which all have different requirements. To provide all these features, a system must offer high performance and parallel processing, as well as efficient communication and synchronization between different, possibly heterogeneous processing units. Figure 1.2 shows the functions of a vision based driver assistance system with the different processing needs of the functions. It consists of functions running well on classical CPUs, as e.g. the feedback loop or standard processing, which require complex computations but work on a small data set. But also of more advanced functions suited for processing on a DSP or GPU based system, as these need to process huge data sets but require less complex computations. Hence, heterogeneous and interconnected systems are needed to efficiently handle the workload.

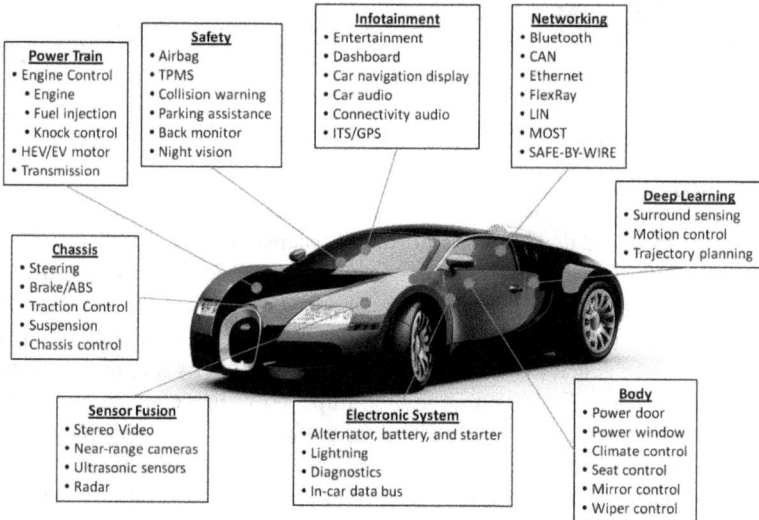

Figure 1.1: *Electronic features used pervasively in automobiles.*

To cope with the increasing complexity of interconnected functions and to reduce the cost and power consumption of a system, multicore systems are used to efficiently integrate different processing units in the same chip. This leads to a transition from many distributed (low performance) ECUs, which require massive wiring and have a high synchronization and communication overhead, over a domain centralized architecture to a software defined vehicle, as shown in Figure 1.3. In a domain centralized architecture or software defined vehicle, high-performance multicore ECUs are used to provide the functionalities, which were previously distributed. And while the domain centralized architecture tries to provide one ECU for each domain for domain-

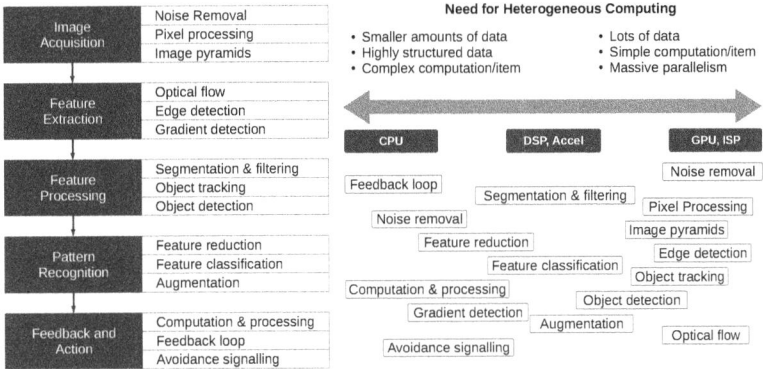

Figure 1.2: *Vehicle computing evolution (based on [1])*

common processing and domain isolation, the software defined approach processes the workload of different domains on the same ECU. Such an approach improves the synchronization and communication between the processing units and hence the performance. At the same time it reduces the isolation properties as functions of different domains with diverse safety requirements are now using the same MPSoCs and network connections, leading to mixed-criticality systems.

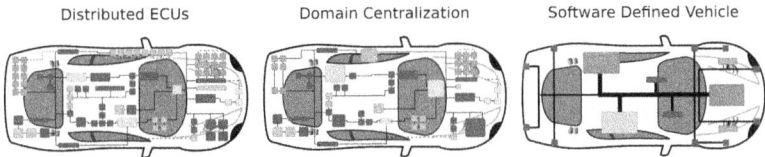

Figure 1.3: *Vehicle computing evolution.*

Especially with the upcoming autonomous driving, the correct functioning of the system must be guaranteed. With the transition of the responsibility from the human to the machine, there will be no driver supervising the decisions and actions of the system, cf. Figure 1.4. Hence, as sketched in Figure 1.5, there will be no driver overtaking in case of errors (as e.g. induced by interferences in mixed criticality systems) and the system must provide a technical fallback. Such technical fallback requires to prove the correct functioning under all cases.

0	1	2	3	4	5
Driver Only	Assistance	Semi-Automation	High Automation	Full Automation	Autonomous

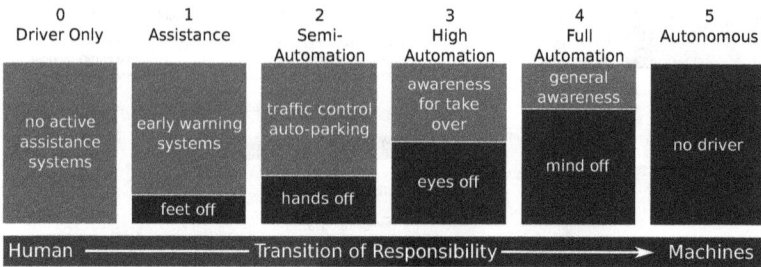

| no active assistance systems | early warning systems | traffic control auto-parking | awareness for take over | general awareness | no driver |
| | feet off | hands off | eyes off | mind off | |

Human ——————————— Transition of Responsibility ————————➤ Machines

Figure 1.4: Transition of responsibility (based on [1]).

Driver is in control and has the responsibility   Fail Safe: Machine is Fallback

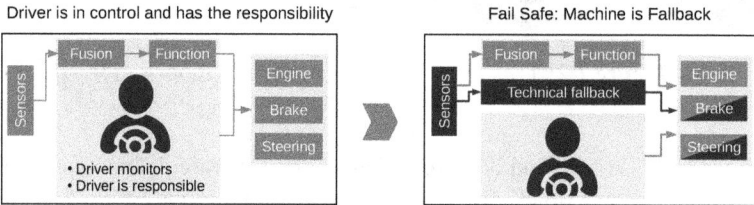

Figure 1.5: Machine as fallback (based on [1]).

Safety standards explicitly mention this problem in context of mixing different criticalities and, for example, require *sufficient independence* (IEC 61508-2, 2010 [15]):

> **7.4.2.3**: Where an E/E/PE safety-related system is to implement both safety and non-safety functions, then all the hardware and software shall be treated as safety-related unless it can be shown that the implementation of the safety and non-safety functions is sufficiently independent (i.e. that the failure of any non-safety-related functions does not cause a dangerous failure of the safety-related functions).

R Sufficient independence of implementation in a mixed-criticality system is established by proving that timing interference or the probability of a dependent failure between the non-safety and safety-related parts is sufficiently low in comparison with the highest safety integrity level associated with the safety functions [15]. While a failure can result from, for example, a fault, wilful timing attack, or wilful memory manipulation and influence timing, data consistency, or other parameters of the system.

Networks-on-chip (NoCs), as a scalable and modular interconnect, are used as a promising solution for MPSoCs, due to their performance, power, and size benefits [31]. In a NoC resources, such as the output ports of the routers, are shared among the different functions and safety classes [126; 216]. Hence, applications of different safety levels will inevitably compete with each other in a NoC for resources (cf. Section 4.1). This resource sharing couples the execution behaviour across cores and, thus, impacts non-functional properties like timing, which are of particular interest in safety-critical environments (as discussed above).

One approach to solve the problem of mixed-criticality is to develop all functions to the highest relevant safety level (cf. IEC 61508-2, 2010 [15]). This leads to higher development costs and lower system utilization. Another approach is to provide *sufficient independence* through quality-of-service (QoS) mechanisms. The challenging part of the latter approach is to efficiently utilize system resources while providing a bounded and feasible interference. This typically leads to a trade-off between providing real-time guarantees for certain applications and performance for the others, as well as the introduced overhead by the quality-of-service mechanisms.

1.2 Standards for Safety

The safety of the public is a major driver of the automotive, railway, industry automation and aviation industry. Based on Part 4 of the IEC 61508, safety can be defined as "freedom from unacceptable risk of physical injury or of damage to the health of people, either directly, or indirectly as a result of damage to property or to the environment" [15]. To ensure the safety of a system there are industrial and research efforts towards standardization of the safety life cycle for electronic products. To reach a safety level and certify or qualify a system, several standards and guidelines must be followed depending on the field of application. There are many national and international organizations, which publish design guidelines and regulations for different domains. Some of these are the International Standards Organization (ISO), the International Electrotechnical Commission (IEC), the Radio Technical Commission for Aeronautics (RTCA) and the Society of Automotive Engineers (SAE). Furthermore, there exist national restrictions by law. Several of the domain specific safety standards are based on the IEC 61508 as sketched in Figure 1.6. This chapter gives a summary of some safety standards relevant for the on-chip network architecture.

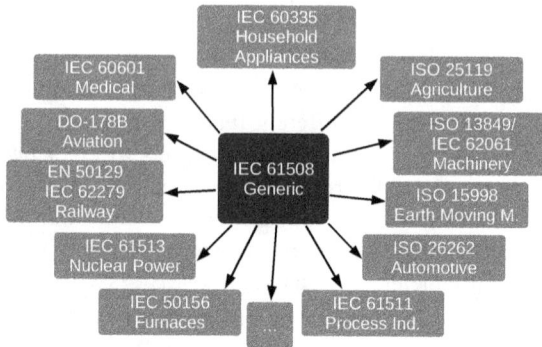

Figure 1.6: IEC 61508 as the root of several safety standards.

The IEC 61508 defines design and verification requirements to establish safety in systems that incorporate electronic/electrical components and their communication [15].

For the avionics, the DO-178B, DO-254, and DO-297 handle respectively the development and validation of software, hardware, and integrated modular avionics [2; 5; 6]. Together these include design considerations on system, hardware, and software level. The ARP-4754 and ARP-4761 describe methods and considerations to get through the certification process of a complex highly-integrated avionics system [13; 14].

Similarly, standards and approaches exist for the development of heavy machinery (IEC 62061), systems for process industries (IEC 61511), railway (IEC 62279), and power plants (IEC 61513).

For the automotive domain so far no necessity for certification exists. However, qualification approaches are used to ensure the correct functioning of the system. The ISO 26262, for example, states that "*If the embedded software has to implement software components of different ASILs, or safety-related and non-safety-related software components, then all of the embedded software shall be treated in accordance with the highest ASIL, unless the software components meet the criteria for coexistence in accordance with ISO 26262-9:2011, Clause 6.*", where Clause 6 proposes "*In the case of the coexistence of sub-elements that have different ASILs assigned or the coexistence of sub-elements that have no ASIL assigned with safety-related ones, it can be beneficial to avoid raising the ASIL for some of them to the ASIL of the element. When determining the ASIL of sub-elements of an element, the rationale for freedom from interference is supported by analyses of dependent failures focused on cascading failures"*. The freedom of interference is later defined as *"absence of cascading failures between two or more elements that could lead to the violation of a safety requirement"*.

Besides the ISO 26262, other standards exists, which can influence the design of an automotive system. The ISO 15005 describes constraints to ensure the safe operation of a road vehicle while it is in motion. This concerns especially the interaction of the user and the vehicle's information and control system [3]. The ISO 16951 handles the prioritized presentation of messages and windows to the user by the vehicle's information and control system [4]. Hence, if such applications use components of an interconnected system, they can influence the design of the network architecture.

By applying these rules to a system-on-chip, the parts of the hardware and runtime environment (RTE), which are always used, must be certified to the highest relevant safety level. For all other components "sufficient independence" must be implemented. Therefore, NoCs, whenever used for communication between safety critical components such as automotive functions, are or will be, depending on the safety-critical domain, the subject of regulation through standards and certification procedures to ensure their correct functioning. In this context, not only the possibly high average performance and low costs play a critical role but also the ability to prove adherence to the safety requirements. This adds another complexity layer to the design process and requires traceability with respect to real-time properties, e.g., application of formal analysis methods such as Real-Time Calculus [213], Network Calculus [134], or Compositional Performance Analysis [99].

1.3 Real-Time Traffic Properties

Modern safety- or mixed-critical embedded systems host heterogeneous applications, with different requirements and behaviour (cf. Section 1.1). This includes applications with different safety-criticality as well as different real-time requirements. An example for the automotive domain is the integration of pedestrian detection in advanced driver assistance systems and entertainment applications. In this sense, criticality can be broken down into at least two orthogonal aspects as shown in Figure 1.7: *safety criticality* and *time criticality*.

For real-time (time-critical) applications the correctness of the system function depends not only on functional but also on temporal aspects. That

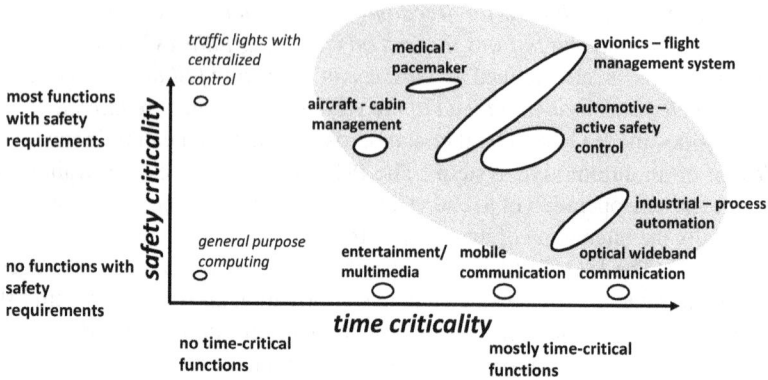

Figure 1.7: *Two dimensions of mixed-criticality [24].*

is, these applications should always finish computations before a given time or receive a certain minimum throughput to ensure correctness or safety. The time by which a specific result must be produced is called *deadline*. Typical examples for such applications without safety requirements are embedded mobile communication (e.g. UMTS or LTE) or entertainment applications.

For purely safety critical systems the integrity of computation needs to be preserved. An example are traffic lights that are controlled by a centralized controller. A failure might lead to catastrophic consequences, e.g., pedestrians or car passengers can be injured or killed. However, it is not important if the (correct) computation result is achieved within milliseconds or seconds. A non-switching traffic light is more acceptable as a wrong state.

Domains with both safety and real-time requirements are of special interest, as they cover many of the important future scenarios like advanced driver assistance system (ADAS) or autonomous driving (cf. Figure 1.7). In these domains, the failure of the function (e.g. violation of a real-time requirement) can have catastrophic effects. While this domain is already challenging, as techniques from real-time and dependability (e.g. predictability) need to be combined, the demands for lower energy consumption and higher efficiency of systems lead to the integration of different domains on the same MPSoC (cf. Section 1.1). In such mixed-critical systems, applications with safety and real-time requirements (e.g. engine control, ADAS) are running together with applications with purely real-time requirements (e.g. entertainment) or no strict requirements at all. Hence, the system must efficiently combine *real-time*, *dependability*, and *high performance* mechanisms.

The timing properties can be further divided into *best-effort*, *throughput bound*, *soft real-time*, and *hard real-time* [71; 114; 142]. Figure 1.8 shows an example for possible timing requirements for some automotive applications.

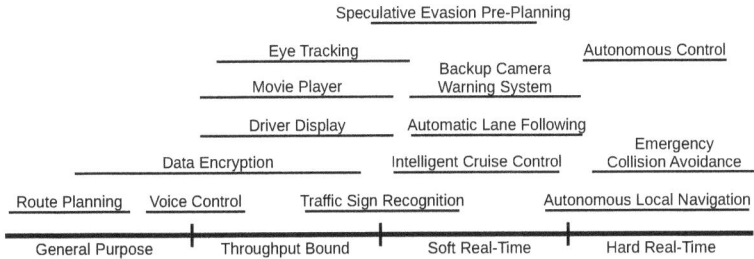

Figure 1.8: Exemplary timing requirements for some automotive applications [71; 114].

Hard real-time applications have firm deadlines, i.e., the utility of the produced result is zero or even renders the whole system as unfeasible when the deadline is crossed. For safety-critical real-time applications, any delay in their execution beyond their deadline, including high latencies of on-chip transmissions, may have severe consequences for the whole system, i.e., fatal failures and prohibitive degradation of service. Hence, these applications are usually not allowed to miss any deadline. That is, for example, the worst-case latency must stay below an assumed upper limit, which is derived from the deadline. For purely hard real-time applications, the functionality does not directly affect user safety but is, for example, important for the user experience. Hence, a deadline violation can cause client loss or substantial financial penalty. To achieve the needed performance for hard real-time applications, the system is typically dimensioned for the worst-case behaviour. Additionally, for safety-critical systems, the correctness of the behaviour, even in the worst-case, must be verified according to safety standards. Due to these requirements the characteristics of safety-critical hard real-time senders and their network traffic are usually well specified and tested and hence known at design time.

Soft real-time applications, on the other hand, may tolerate occasional deadline misses. The main difference, when compared to hard real-time senders, is that these applications are rarely required to rigorously meet all their deadlines, i.e., the produced results have some utility after the deadline or are not affecting the safety (e.g. can safely be discarded) [142; 201]. An example are control algorithms based on a feedback loop or video analysis for night vision in a car. The algorithms in such systems may tolerate a limited number of cases when instead of new sampling data old values are used. Thus, it can compensate occasional deadline misses without any severe consequences for the system [193; 218; 239].

Similar to soft real-time, there are throughput bound applications, which must comply to overall real-time performance objectives in terms of a minimum achieved throughput over a given period. Again, these applications are rarely required to rigorously meet all their deadlines, i.e., the short term throughput can underrun the required throughput as long as the long-term throughput is acceptable. For instance, video streaming done as a part of an infotainment function in a car does not influence vehicle safety, but video frames must still arrive with a certain latency to prevent quality drops and glitches. Still, for a producer of an infotainment system the quality of user experience may play a critical role in the market success of a product. Consequently, a producer may accept sporadic drop of the video quality but may lose clients whenever it happens too often.

The last category of general purpose or best-effort (BE) does not have strict temporal requirements, e.g, a deadline miss will not endanger system safety. However, this typically also means that such application are less tested with respect to temporal properties and only designed targeting average performance metrics. For example, the frequency and size of accesses to the (on chip) network are not known. Still, the system should provide sufficient resources to process such applications (e.g. be work conserving), as they can, for example, be used to increase the long-term efficiency or user experience (e.g. diagnosis functions or route planning). Hence, achieving high performance and low latency is a common design goal, as long as all guarantees for safety- or time-critical applications can be delivered.

1.4 Requirements of Safety-critical Embedded Systems

As shown in the sections before, the communication of safety-related data must be protected at run-time against effects of faults, which may lead to failures of the system. These faults include transient faults, physical damage as well as lack of sufficient independence between tasks (e.g. timing interference or data corruption). For example, the ISO 26262 provides a list of faults, presented in Table 1.1 regarding the exchange of information, which must be considered in case of an interconnect for certification purposes. An end-to-end protection defines a set of mechanisms, which avoid these faults or allow a reliable detection and appropriate countermeasures.

Some of these faults directly relate to real-time metrics for the on-chip interconnect, e.g., a *delay of information* or *blocking access to the communication channel*. Others relate to consistency and the protection of packets done directly in the interconnect. For example, without a proper flow control in the network, packets might be dropped or overwritten leading to corruption of information. Other faults, although not directly related to the temporal metrics, can influence the predictability indirectly. Transient errors, malfunctioning, or malicious senders, for example, can introduce uncertainty and dynamics to the system, e.g., sporadic overloads due to re-transmissions or babbling idiots. Such dynamics hinder predictability or even render it impossible.

These faults can be detected and partly avoided in the software layer of a system. For example, AUTOSAR provides several mechanisms to cope with these faults (cf. Table 1.2), as e.g., *CRC, Data ID, Counter, Regular transmission + timeout monitoring* [10; 77]. The *Data ID* is a unique identifier to verify the identity of each transmitted (safety-related) data element. The *Counter* is a simple counter that is incremented on every send request. It can be used to implement an *alive counter* and a *sequence counter*. For the sequence counter, the value is checked at receiver side for correct incrementation, while for the *alive counter* it is only checked whether is changes at all. Based on these mechanisms receiver communication and sender acknowledgement timeouts can be implemented. For this, a receiver is executed independently of the data transmission (e.g. periodic activation and checking for new data) and checks the validity of the received data (based on CRC, counter, and Data ID). With this, a wrong counter detects a duplication of previous data, loss of communication, or timeouts. Table 1.2 shows the fault coverage of different mechanisms. These mechanisms can be realized in software (cf. AUTOSAR E2E Protocol Specification [10]), hardware, or a hybrid solution. To increase the efficiency of a system, the hardware can provide support to detect and avoid such faults, e.g., hardware CRC checking or quality of service (QoS) mechanisms to limit interference. The latter are of special interest, as such mechanisms change the timing and behaviour of the system. Additionally, they can drastically degrade the performance, utilization, or adaptability of a system. For the other mechanisms (and faults), the influence on the system design and behaviour can be straightforwardly derived.

Table 1.1: *Summary of communication faults in the design of the automotive systems according to ISO 26262 [7; 10].*

Fault Type	Description
Repetition of information	A type of communication fault, where information is received more than once.
Loss of information	A type of communication fault, where information or parts of information are removed from a stream of transmitted information.
Delay of information	A type of communication fault, where information is received later than expected.
Insertion of information	A type of communication fault, where additional information is inserted into a stream of transmitted information.
Masquerading	A type of communication fault, where non-authentic information is accepted as authentic information by a receiver.
Incorrect addressing	A type of communication fault, where information is accepted from an incorrect sender or by an incorrect receiver.
Incorrect sequence of information	A type of communication fault, where information is accepted from an incorrect sender or by an incorrect receiver.
Corruption of information	A type of communication fault, which changes information.
Asymmetric information from sender to multiple receivers	A type of communication fault, where receivers do receive different information from the same sender.
Information from a sender received by only a subset of the receivers	A type of communication fault, where some receivers do not receive the information.
Blocking access to a communication channel	A type of communication fault, where the access to a communication channel is blocked.

Table 1.2: *Fault detection coverage for different mechanisms.*

	Counter	Data ID	CRC	Transmission on regular bases and timeout monitoring	QoS
Repetition of information	x	—	—	—	—
Loss of information	x	—	—	—	—
Delay of information	x	—	—	x	x
Insertion of information	x	x	x	—	—
Masquerading	—	x	x	—	—
Incorrect addressing	—	x	—	—	—
Incorrect sequence of information	x	—	—	—	—
Corruption of information	—	—	x	—	—
Asymmetric information from sender to multiple receivers	—	—	x	—	—
Information from a sender received by only a subset of the receivers	x	—	—	—	—
Blocking access to a communication channel	x	—	—	x	x

As discussed in Section 1.1, NoCs are foreseen as a communication backbone for large systems-on-chip (SoCs) integrating different ECUs. As a result of such integration, it can happen that different traffic classes, i.e., hard real-time tasks, soft-real time, throughput bound, and best-effort (BE), share the SoC resources. This causes co-dependencies between applications running on different cores, what may endanger safety. Unpredictable and bursty accesses from BE senders may lead to contention in network buffers. In on-chip interconnects without appropriate QoS mechanisms transmissions are scheduled as soon as they arrive, and all traffic receive equal treatment. This leads to the possibility of an unbounded timing interference, which may lead to missed deadlines for real-time traffic. Hence, NoCs for future safety-critical systems need to provide QoS mechanisms.

NoC architectures are judged by *performance* (e.g. latency, throughput, utilization), *cost* (e.g. design effort, HW overhead), *predictability* (e.g. formal analysis; guarantees on performance metrics), and *flexibility/adaptability* (e.g. adapt to system internal and external changing conditions; re-use for different use cases) [24; 54; 69]. The challenge in providing mechanisms for these is that they can be contradictory. For example, an arbitration based on a static time-division multiplexing (TDM) achieves a high predictability.

Under certain conditions, it can even achieve a fair performance, e.g., when the TDM slots can be fully utilized. However, the design is usually developed according to the worst-case behaviour, leading to unused slots during normal operation of the system. Hence, the design is not work-conserving leading to a low utilization and degradation of performance. Additionally, a TDM approach can typically not easily adapt to changing conditions. Summarizing, the most important requirements of a NoC architecture are:

- efficient support of different traffic types
  - sufficient independence (limited interference between applications)
  - guaranteed worst-case performance for real-time applications (predictability)
  - as good as possible actual-case performance for non-real-time applications (e.g. high utilization, work conserving, no "second class citizens")
- low cost
  - low design effort
  - low hardware overhead
  - reusability: allow the same architecture to be used for different usecases/domains
  - allow efficient verification (bounds on interference)
- flexibility
  - allow to adapt to system internal and external changing conditions
  - reusability: allow the same architecture to be used for different usecases/domains
  - safety and real-time "as a feature"

1.5 Research Objective and Contribution

For safety-critical systems, a major goal is the avoidance of hazards. For this, safety-critical systems are qualified or even certified to prove the correct functionality under all possible cases. A predictable behaviour can help to ease the qualification process (e.g. analysis) of the system. Thus, achieving a predictable behaviour is an important goal for these systems. For the interconnect (e.g. the NoC) design, this means that providing predictable resource sharing between concurrent transmissions is a major driver. However, the support for temporal properties should not cancel out benefits resulting from the application of NoCs, e.g., high efficiency, scalability, flexibility, and low production costs. Hence, these other design goals shall also be reached, to reduce cost, increase efficiency, and market competition.

To achieve the required predictability, designers have two classes of solutions: isolation (QoS mechanisms) and (formal) analysis. For mixed-criticality systems, isolation and analysis approaches must be combined to efficiently achieve the desired predictability. Isolation techniques are used to bound the interference between different application classes. And analysis can then be applied verifying the real-time applications and *sufficient isolation* properties.

For safety-critical systems, state-of-the-art approaches for isolation have adverse impacts on system performance (especially for the non safety-critical applications) and on the flexibility of the system. These mechanisms usually prioritize hard real-time senders over best-effort traffic and soft real-time traffic. Hence, BE traffic suffers from high latencies although time critical traffic has no to little benefit from a reduced latency [201]. Moreover, state-of-the-art approaches apply a worst-case dimensioning, which frequently leads to resource over-provisioning. However, during regular work of the system such extreme conditions may rarely occur. This results in a significant drop of average utilization, i.e., underutilized resources. Consequently, an efficient co-execution of mixed-critical applications is still an open research question with possibly high engineering and economic impact.

The goal of this work is to develop a NoC architecture that provides timing/isolation guarantees (for safety-critical applications), high performance (i.e. less adverse impacts for BE), and flexibility (application dynamics, design dynamics, errors, updates, usage in different use cases). Additionally, the work shall provide the mean to analyse the architecture to prove its real-time (and performance) capabilities.

The contribution of this thesis to address the aforementioned research objectives is twofold: We present a modelling and analysis framework for NoCs that accounts for backpressure (i.e. full buffers in network routers delaying the progress of network packets). This framework enables the evaluation of performance and reliability early at design time. Hence, the designer can assess multiple design decisions and trade-offs (such as area, voltage, reliability, performance) by using abstract models and formal approaches. State-of-the-art analysis approaches typically cannot handle backpressure. Hence, the occurrence of backpressure must be avoided in the system. This can be reached by providing sufficiently sized buffers to avoid buffer overflow or by restricting the interconnect access rates. The first case leads to a resource over-provisioning and high hardware overhead due to the buffer sizes. This results from the fact, that the buffers must be configured to cover the worst-case behaviour of the system. Additionally, analysis approaches apply some conservatism amplifying this effect. The seconds case leads to a decreased performance of the system. In such a design, the traffic source is only allowed to send a certain amount of data to the interconnect by using, for example, rate limiters. To avoid buffer overflow, the rate limiters must again be designed according to the worst-case. Hence, they typically slow down the applications too often during normal operation, leading to decreased performance. With the new backpressure-aware analysis framework, buffer overflow is covered by the analysis avoiding deep buffers or restrictive rate-limiters. Hence, such an analysis allows a more efficient design for the interconnect and helps to avoid the conservative over-provisioning of buffer space.

Furthermore, we provide a QoS-aware architecture that provides safety support, high performance, and flexibility. For this, we investigate different QoS mechanisms for a NoC. This includes two hardware based mechanisms (i.e. QoS-aware router designs) and a HW/SW co-design approach. The latter uses a *control layer* for a safe but flexible and high-performance admission control and NoC resource management. While the general benefits of the NoC resource management are already explored in [122; 125; 223], this thesis provides an architecture supporting the control layer of the NoC resource management. This support allows to increase the benefits of the NoC resource management and to reduce its overhead w.r.t. to area and performance.

To address the previously described research objectives, this thesis is structured as follows:

Chapter 2 provides an introduction to the basics of NoCs and an overview of some existing NoC architectures.

Chapter 3 provides the backpressure-aware analysis to formally verify real-time constraints of a NoC.

Chapter 4 investigates different QoS mechanisms for a NoC. First, we investigate hardware mechanisms for guaranteed latency and throughput (i.e. QoS-aware router designs). This is followed by an overview on a control-layer centric QoS mechanism, which can simplify the network routers and increase the flexibility of an architecture.

Chapter 5 provides a NoC architecture that supports the control layer for QoS to increase its benefits and shows its general applicability to NoC designs.

Chapter 6 provides an evaluation of the proposed architecture.

Chapter 7 summarizes this work and draws a conclusion, including possible directions for future work.

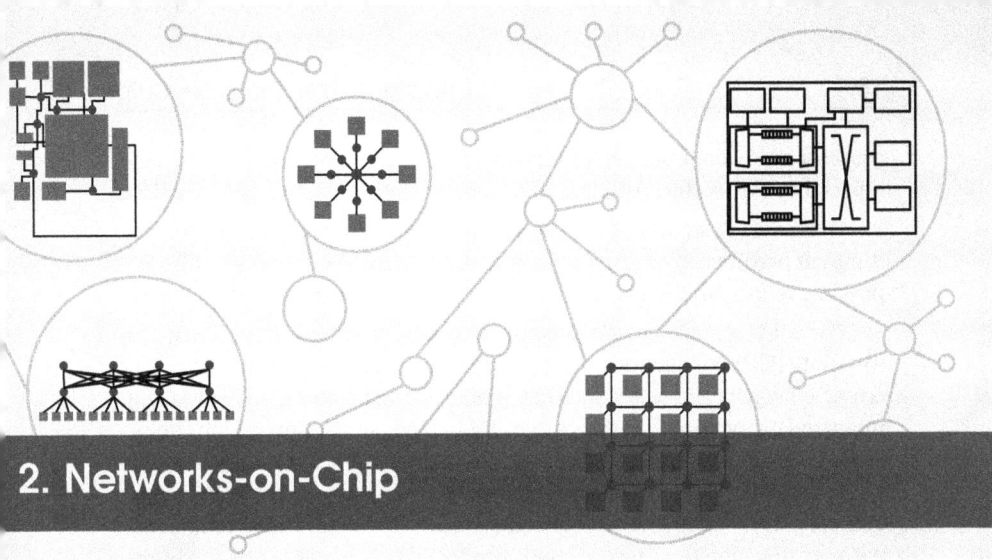

2. Networks-on-Chip

This chapter provides an introduction to the basics of networks-on-chip (NoCs) and an overview of some existing NoC architectures. NoCs, as a modular interconnect, are used as a promising solution for multiprocessor systems on chip (MPSoCs) (cf. Section 1.1). A NoC has three major building blocks: network interfaces (NIs), routers, and network links. A dedicated NoC architecture is formed by interconnecting these elements in different configurations (forming a topology) and choosing certain functionalities and implementations for each element.

2.1 Network-on-Chip Basics

There exist several architectures for the on-chip interconnect of systems-on-chip (SoCs). Some commonly used architectures are the *shared bus*, the *crossbar*, and the *network-on-chip (NoC)*.

The shared bus is widely used in small or low performance commercial SoCs. In a bus-based system the components are connected to the same shared medium (the bus) and divided into different groups, namely master and slave devices. A bus arbiter grants a master access to the bus with respect to an arbitration protocol. Bus-based architectures are very simple to implement and have a low area cost. However, the simplicity of this architecture results in some disadvantages. One disadvantage is the absence of scalability, since there is a high competition for bus cycles with an increasing number of devices (e.g. processors). The shared medium between all components also causes a low overall throughput. Examples for bus architectures are amongst others the AMBA Bus, Avalon, CoreConnect, Marble, STBus, and Wishbone.

Bus-based systems can also be hierarchically structured or divided into different partitions by the usage of *gateways*. In different layers or tiles it is possible to use different buses or even interconnect architectures.

For a low number of components, a crossbar switch or dedicated point-to-point (P2P) connections offer a second interconnect architecture. The crossbar switch can be used to offer a point-to-point connection between any input and every output of the switch. Dedicated point-to-point links do not require a switch in between, but multiple interfaces at each component. For a system with four components (e.g. cores), for example, these architectures can offer two independent connections between disjoint communication partners (e.g. core0 with core1 and core2 with core3). This offers a higher overall throughput compared to the shared bus, since concurrent connections are possible. Nevertheless, an increasing number of channels in the crossbar or needed interfaces for dedicated links causes a raise in complexity. This limits the amount of components that can be interconnected efficiently in such fashion. And while some designs for crossbars with a high number of ports exist, the complexity of physical routing of wires gets complex [169; 170].

Another approach is the usage of networks-on-chip (NoCs). Here the components are connected to a network of switches (also called routers). Figure 2.1 shows a conceptual overview of a NoC based system. In a NoC, packets are used to exchange data between different nodes. Between distinct switch pairs concurrent connections to send the packets are possible. The computing blocks are commonly connected through a network interface (also called network interface unit (NIU)) to the routing elements. The network interfaces (NIs) naturally offer the possibility for isolation between a component and the network (or remainder of the system). In such a switched network several possible routes between two components can exist, which offers redundancy for the communication channels and can be used to improve the reliability of the connection and for load balancing.

Based on analytical estimations and measurements on an FPGA prototype, Ogras and Marculescu identified the following properties of the different architectures [164]:

1. The performance and scalability of NoC-based implementations is very close to that of the point-to-point (P2P) for the same application, while bus-based implementations scale much more poorly.

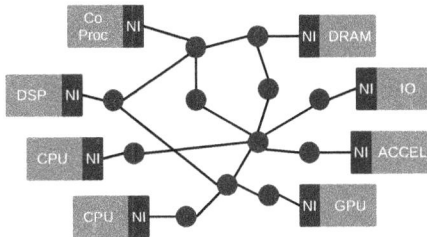

Figure 2.1: Conceptual view of a network-on-chip (NoC).

2. In terms of area, the NoC scales as well as the bus. The P2P implementation does not scale well due to the overhead involved in redesigning the interfaces. This also allows the integration of new cores to an existing design to be much easier in NoC-based designs.

3. The energy consumption of NoC-based architectures is smaller and scales better than that of P2P or bus-based designs.

In summary, bus-based architectures have a low design effort and area overhead. But they scale poorly with respect to performance and energy consumption. Point-to-point architectures can deliver a high and scalable performance. However, they scale poorly in the overall design effort, energy consumption, and area overhead. Compared to this, NoC architectures scale better in terms of overall design effort, area overhead, energy consumption, and performance. Hence, NoC architectures are a reasonable choice for future embedded systems.

Argarwal et al. [22] and Bjerregaard and Mahadevan [36] present an overview of the main properties of network-on-chip design. When designing a NoC several aspects, e.g. the *topology*, the *routing protocol*, the *switching technique*, and the *flow control*, have to be defined. All these aspects have a great impact on the behaviour and predictability of the system under design. The most important aspects are summarized in the following of this section.

2.1.1 Topology

The topology of a NoC defines how the network nodes are connected to each other, i.e., the shape of the network. This impacts directly the system cost and performance, as well as the reliability of the network. Figure 2.2 shows some commonly used topologies. There are several parameters defined by the topology. The most important are the *diameter*, *link capacity*, and *node degree*.

The node degree defines the number of input and output ports of the network. This can be taken as a measure for the input/output complexity of a node. If all nodes have the same degree, the network is called *regular*, otherwise *irregular*. The node degree can be constant or varied according to the network size. Typically, a high node degree reduces the average path length but also increases the complexity, where a smaller node degree requires less hardware cost on links. In many cases there is a constraint on the node degree, which results from the number of direct neighbours of a node. For topological characteristics, a small and fixed node degree is more effective.

The diameter describes the distance between nodes, i.e., the maximum number of links (or maximum hop count) between any two nodes taking the shortest path. Each node travelled by a packet introduces a delay and hence the overall delay increases with the maximum hop count. A small diameter can help to achieve low latencies and to ease routing (e.g. as fewer decisions need to be made).

The link complexity describes the number of links. As a network scales, the link complexity increases. More links in a network can help to reduce the diameter and provide better communication between nodes. However, a higher link complexity increases the hardware and area overhead.

Related to this, the bisection width and bisection bandwidth of a network are metrics for the complexity and performance. The bisection width describes the number of links between two sub-networks, if the topology is divided into two networks with approximate equal size. A higher bisection width then provides more paths between the two sub-networks and thus improves the overall performance of the network. The bisection bandwidth then describes the throughput that can be achieved between these two sub-networks.

In the *mesh* topology nodes form a 2D-array/grid, where each inner node is connected to the four adjacent routers and the routers at the edges or borders have two or three connections. A simple mesh network has a node degree of 5, but may vary according to the type of the mesh network. The *torus* topology extends the mesh by wraparound link on the edges. Hence, each node (also the ones at the borders and edges) has four connections to neighbouring routers. This enables a better path diversity than in the mesh and reduces the hop count. In an *irregular mesh-based* architecture, nodes are placed in a grid. However, a component might span multiple grid elements. Instead of connecting such component to multiple network routers, the unnecessary routers are removed from the grid, leading to an irregular

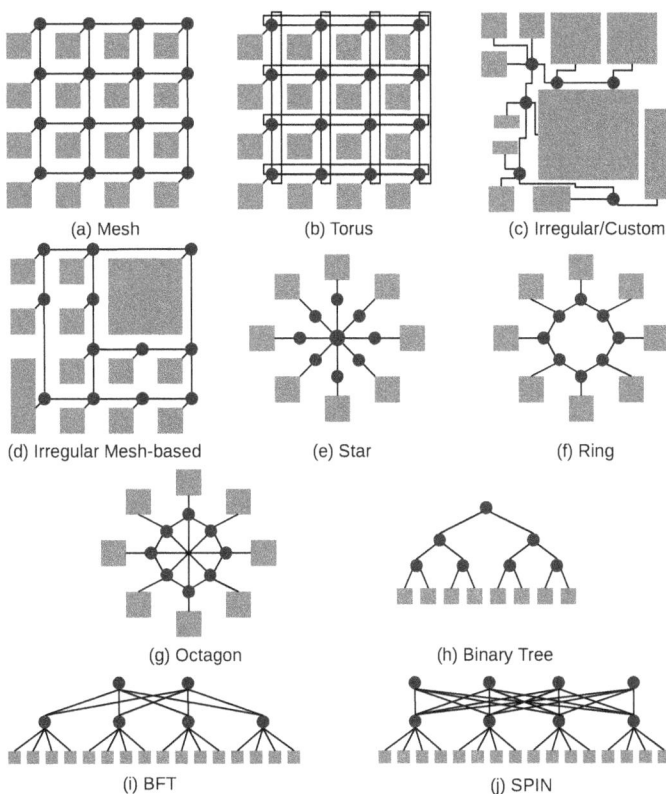

(a) Mesh   (b) Torus   (c) Irregular/Custom

(d) Irregular Mesh-based   (e) Star   (f) Ring

(g) Octagon   (h) Binary Tree

(i) BFT   (j) SPIN

Figure 2.2: *Examples for different network topologies.*

network. In the *star* topology a central switch connects all resources. In the *ring* topology every node is connected to two neighbouring nodes forming a ring. A transmission through the network passes each of the nodes in the ring until it reached the destination node. The ring topology has a fixed node degree of three. The *octagon* is an extension of the ring, consisting of eight nodes and 12 bidirectional links. Additional to the links in the ring, there exist direct links between adjacent nodes. In a *tree* topology, a central switch is used as the root of a tree and is connected to one or more nodes of a lower hierarchy. The resources are then connected at the leafs of the tree. One example is the butterfly fat tree (BFT). If each node has a specific fixed number of nodes in the lower level it is connected to, the tree has a symmetrical hierarchy. The *Scalable, Programmable, Integrated Network (SPIN)* architecture is similar to a butterfly fat tree topology. Here, each node in a level has the same number of ports to the higher and lower level.

2.1.2 Routing

The routing defines the strategy to select a path from the sender to the receiver in a network [54]. The routing protocol is crucial for the safety and predictability of the system. If dynamic or adaptive routing protocols are used, the traversal time of a packet through the network is unknown because it dynamically changes depending on the system state. Without evidence that this time can be bounded, the behaviour of the whole system gets dynamic and unpredictable. Additionally, it influences the load distribution across the network, the overall throughput and performance, and the connection path length, influencing latency and the amount of possible points of failure. The routing protocol determines also how the system will behave when errors on a link occur.

Several aspects can allow classification of the different routing protocols. The most important aspect is the adaptability. The routing can be deterministic or adaptive. A deterministic routing always uses the same path through the network and the path is statically defined during the design time along with the choice of the routing protocol like dimension ordered routing (e.g. XY-routing). This determinism avoids deadlocks and leads to a better analysability of the network. Adaptive routing protocols, on the other hand, use the current network state to dynamically determine an optimal route. In such a way, they allow better load distribution and network utilization but usually are more complex and harder to analyse. Therefore, adaptive routing is not as predictable and reliable as deterministic routing. Detailed overviews of different routing algorithms have been presented, for example, by Rantala et al. [187] and Mello et al. [148].

Besides the general routing protocol, routing can be implemented at the source (source routing) or distributed in the network. For source routing the sending node (e.g. its network interface) decides on the path through the network. This path is typically stored directly in the packet header. Each router then just selects the correct output port based in this information. This approach can simplify the router architecture and lead to a lower latency [69; 106; 155]. The route can be stored in various encodings in the packet header. For example, each hop can be encoded directly or a run length encoding can be used. As source routing adds information to the header, source routing schemes typically introduce overhead to the packet and may reduce the effective utilization. The overhead depends on the encoding, the size of the network, and the size of a packet (e.g. the payload to header ratio).

For distributed routing protocols only the address of the destination node is stored in the packet header. Each router then computes based on this address the correct output port. This computation can induce an additional delay in each router (compared to source routing). As only the destination address has to be stored in the header, this approach induces less overhead to the packet. However, as not the full path is stored in the header, the sender of a message cannot directly be identified. Hence, if knowledge of the identity of the sender is required, a *sender ID* needs to be stored in the packet header or payload.

Figure 2.3 shows the required number of bits in the packet header for a symmetric mesh topology with a single node per router, where in addition to the destination address also the address of the sender is encoded in the packet header. For source routing scheme, the sender ID can be extracted from the route while for distributed routing an additional sender ID field in the packet header is needed. The figure shows that distributed routing scales well with the NoC size. However, for distributed routing each router has to compute the desired output port before a packet can be processed. Hence, distributed routing typically requires one pipeline stage more than the other schemes. This additional pipeline stage can be avoided by speculation, which in turn leads to more complex router designs. The basic source routing, where each hop is encoded in the header, scales worse with the NoC size. For the run length encoding two variants are shown. The first variant stores up to four directions where a single direction can cross the whole NoC in one direction. That is, the length field of the run length encoding covers the maximum hop count for a single direction. The second variant uses a reduced length field, such that a single run can only cross half of the direction. Hence, at least two runs are needed to cover the maximum hop count for a single direction. Doing this, the size of length field and the counter for the current run can be reduced. The different run length encodings scale well, depending on the number of used directions and the length of a single run. However, for small NoCs there is an overhead compared to the other schemes.

2.1.3 Switching

The switching technique defines how connections are established in the network. It specifies how a router removes a packet from an input port and places it to an output port and thus influences how the network allocates channels and buffers to the packets.

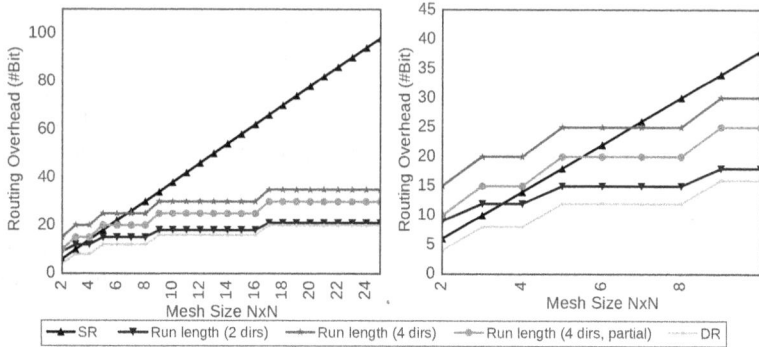

Figure 2.3: *Routing overhead of different encoding schemes.*

Switching techniques can be divided into circuit switching and packet switching approaches. Circuit switching approaches allocate the entire source-destination path before the data is sent. With an allocated path there is no need for routing information in the header of the packet stream. The routing information is, for example, stored in a special configuration message reserving the whole path. Circuit switching can be pipelined along the path through the network. The pipelining and the absence of routing information lead to a high performance for continuous streams. However, for short messages, the overhead through the path reservation limits the performance. The alternative to circuit switching is to divide the data into packets that can be transmitted independently through the network and therefore require not to reserve the whole path at once. Packet switching techniques can be divided in *Store and Forward*, *Wormhole*, and *Virtual Cut Through* switching. In *Store and Forward* (S&F) switching a complete packet is stored at each stage in the network before the next link is switched. This decouples the links completely from each other. The storing of the packets at each stage induces an additional latency on each stage to each packet and requires enough buffer space in the switches to store whole packets. The buffer size and the storing time on each switch can be reduced with the usage of smaller packets. Another way to reduce the induced latency caused by storing the whole packet is to forward parts of it, before the complete packet has arrived, as done in cut-through packet switching. In this approach, the header of a packet can be forwarded immediately to the next link while the payload is still on the current link (or previous router). On each stage the packets are processed in a pipeline, which makes this approach very efficient for multi-stage networks. There are two different versions of the cut through switching: the *Wormhole* and *Virtual Cut Through* (VCT) switching. In VCT switching the flow control and buffer allocation is done on the packet level, i.e., a packet is completely buffered in a switch if the header is blocked. Therefore, the router requires sufficient buffer size to store the complete packet at every switch like in the case of S&F switching. In wormhole switching the packets are split into smaller *flow control units* (flits). These flits are then transported independently through the network and a packet is represented as a stream of flits. If a flit is blocked, all following flits are blocked too, but possibly on different switches. Because of that, buffers used for wormhole switching can be smaller, capable of storing data in the size of a flit. This reduces the cost but leads to an earlier congestion in the network when not using special mechanism to avoid congestion (like *virtual channels*). A survey of wormhole switching techniques in comparison with other techniques was presented by Al-Tawil et al. [21].

Usually the selection of a particular technique strongly depends on the needs of a platform. For example, a store-and-forward technique was used in Nostrum [151] network while designers of Æthereal [81] and Mango [37] decided to use combinational techniques depending on their different needs.

2.1.4 Virtual Channels

Virtual Channels (VCs) are used to split a physical link into multiple virtual links. For the implementation of a VC on a link the buffer on the link (e.g. in the ingress or egress port of the switches) has to be extended to include multiple (virtual) buffers, one for each VC. By providing multiple buffers for each link, the allocation of buffers is decoupled from allocating of links. Virtual channels increase the area needed for the switch but introduce some advantages. One advantage is the reduction of the deadlock and livelock problematic. The VCs introduce more output paths per link, which reduces the probability for streams to compete for the same buffer. Additionally, the VCs introduce an isolation of the different concurrent streams in the network. A VC can also be used to offer support for quality of service (QoS) for parts of the communication. For this, a VC can be reserved for high priority traffic and be privileged in all switches. Virtual channels also allow packets to pass blocked packets or streams and hence links can be used that otherwise would have been idling. This increased utilization of the links then can lead to a higher overall performance of the network.

Table 2.1: Summary of channel and buffer allocation schemes.

Technique	Channel	Buffer	Description
S&F	Packet	Packet	Head flit must wait for the arrival of the entire packet before proceeding on next link.
VCT	Packet	Packet	Head flit can begin traversal on next link before tail flit arrives at current node.
Wormhole	Packet	Flit	Channel allocation on packet level.
VC	Flit	Flit	Can interleave flits of different packets on links (if they use different VCs). Can be combined with the three others.

Virtual channels can be combined with the three switching techniques mentioned before. Table 2.1 provides a summary for the different buffer and channel allocation schemes. In the table, the *buffer* column denotes how buffers are allocated. That is, if there must be sufficient buffer space for a packet or flit at the next node before a flit/packet can be transmitted. The *channel* column then denotes how the link/channel between two nodes is allocated to transmissions. That is, if a packet needs to be fully transmitted before a packet from another transmission can be scheduled or if different packets can be interleaved.

2.1.5 Flow Control

The flow control determines how the downstream node communicates forwarding availability to the upstream node (i.e. the buffer management). It is needed for non-circuit switched networks to deal with congestion at the network buffers. Flow control is connected to the switching strategy. Basically, the switching strategy defines the size of the data chunks and needed buffer space before data may proceed, while the flow control monitors the current available buffer space, such that data is only forwarded when the needed space is free or delayed until the buffer is free. The choice of the flow control technique influences the fraction of the ideal bandwidth that can be reached and the predictability of the timing in the network. For example, a poor designed control flow technique might lead to idling resources when not necessary and therefore a low utilization.

For buffered flow control techniques, it is important to control the availability of the buffers during the communication in order not to lose or overwrite information and thus guarantee reliability of the transmission. Hence, a network node communicates the availability of its buffers to the preceding node. If no buffer space is available, a packet backlog can occur leading to blocking and possibly to a propagation of blocking. This effect is called *backpressure*. Common implementations for buffered flow control are:

- *credit-based flow control*: For the sending node an amount of credits is defined and decreased when sending a flit. When the receiver node processed a flit, it sends a credit signal back to the sender to increase the available credits.

- *ON/OFF flow control* (also called *STALL-GO*): Is a simple technique using one control bit as a flag, signalizing if a sending node is permitted to send the data (ON) or not (OFF).

- *ACK/NACK protocol*: The sending node sends a flit and buffers it until an ACK or NACK signal is received from the receiving node. When an ACK is received, the next flit is sent and when an NACK is received, the old flit is sent again.

2.1.6 Baseline Switch Architecture

This section presents a basic architecture of a NoC switch. It is based on the one presented by Dally [54]. It constitutes a common selection for the parameters introduced above.

Figure 2.4 presents a simple block diagram of the switch architecture using wormhole switching and virtual channels. It consists of input and output modules, which are connected by a crossbar, and control modules. The control modules comprise route computation, VC allocator, switch arbiter, and credit counters. Together they control the transfer of data.

An input module has several sub-modules: a demultiplexer, buffers, and a multiplexer. As the presented switch uses virtual channels, it has several FIFO buffers per input module (one FIFO buffer for each VC) to prevent head-of-line blocking between different VCs. The FIFO buffers are organized in fix-sized flow control digits (flits), i.e., they can store one or multiple flits. Data arrives from the input channel and the input demultiplexer assigns the arriving flit to the corresponding VC buffer (based in the VC in the flit header). To reduce the complexity of the crossbar, the FIFO buffers of an input module are connected through a multiplexer to the crossbar. Hence,

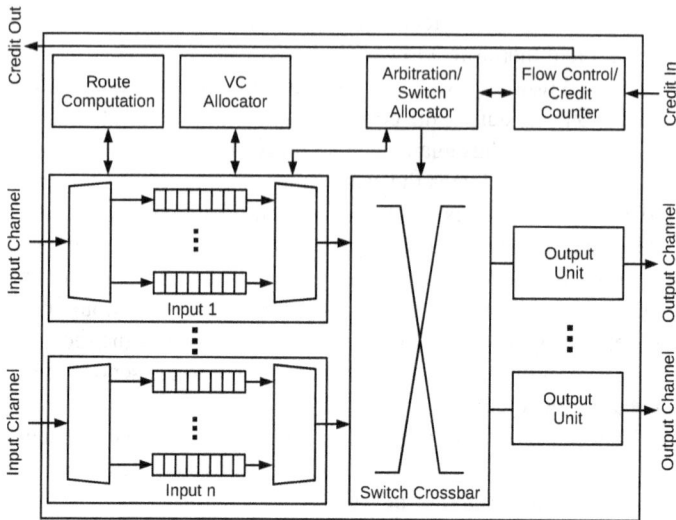

Figure 2.4: *Baseline switch architecture.*

an input module can only send a single flit per cycle through crossbar to an output module. The output module then is responsible for forwarding the flit to the next router. It can contain buffers for output-buffering, but the presented design uses only input-buffering.

Each control module can be implemented as a centralized module or decentralized. The route computation module computes the output port of a packet. Hence, it is invoked once per packet (i.e. on the arrival of the head-flit). There can be a single route computation module per switch to save area or one for each input module. The latter approach has the benefit of computing the output port for several input ports in parallel. With the known output port, the VC allocator selects an output VC for the packet. This module is again invoked once per packet. There can be a global VC allocator on the switch or one for each output port. With the known output port and VC, the switch allocator selects the time slot on the crossbar to forward a flit to the output. The switch allocator is invoked for each flit. As there might be multiple input modules or VC requesting for the same output port, the access to the crossbar and output ports also needs an arbitration. The arbitration can be implemented in a centralized or distributed way. For a centralized implementation, all requests and status information are handled by a central

In 1	Out 1		In 1	Out 1		In 1	Out 1
In 2	Out 2		In 2	Out 2		In 2	Out 2
In 3	Out 3		In 3	Out 3		In 3	Out 3
In 4	Out 4		In 4	Out 4		In 4	Out 4

Request   Grant   Acknowledge

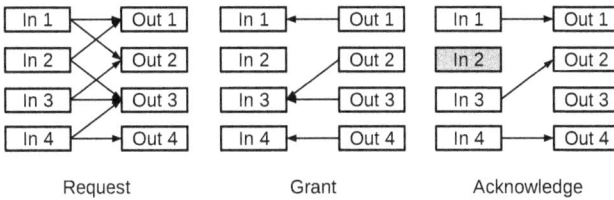

Figure 2.5: *Three phase arbitration between input and output ports.*

arbiter. For the distributed implementation, the arbiter can be distributed across the input or output modules. As the arbitration is performed for each flit and there can be a high number of requests (e.g. one for each input VC), the arbitration is typically implemented hierarchically. That is, for example, the arbitration is first performed at the output module, deciding which input is allowed to use the output port. Then, in a second arbitration step, the input module decides on the VC that is forwarded. This leads to three phase arbitration as shown in Figure 2.5. In the figure four input ports are sending requests to possibly multiple different output ports (e.g. when in an input module the VCs requests for different output ports). In a second phase, the output arbiter decides on the input port that can use the output port, for example, based on a round-robin policy. It then sends a grant to the input port that won the arbitration. As an input module might receive multiple grants, it locally arbitrates between the multiple VCs that received a grant and sends an acknowledge to the output port.

Processing of a Packet

With the architecture described above, the transfer of a packet in a switch is performed in multiple stages: upon the reception of a packet, or more precisely of its head flit, the route computation module decides to which output port the packet is forwarded based on destination information stored in the head flit. In the second step, the packet requests a virtual channel (i.e. a FIFO) in the next switch (along the path of the packet) at the VC access controller. For source routing schemes these steps can be done in parallel, where in the route computation step the output port for the next hop is calculated. With the known VC, the packet can proceed and request access to the output port (and crossbar) at the switch arbiter. This is only done if there is sufficient buffer space in the downstream VC, which is determined using the credit counter module. If this is the case, the switch arbiter decides when a specific flit (e.g. input module and VC) may use the crossbar and hence be transferred to the downstream switch. Upon the successful transferral of a flit, the switch arbiter asserts the credit out signal to let the upstream switch know that space became available in the corresponding VC. While route computation and VC allocation only happen for the head flit, switch arbitration is done for every flit of a packet. Once the tail flit has been forwarded, the allocation of the VC is released.

With this, we can differentiate four typical pipeline stages as shown in Figure 2.6. During the *route computation* (RC) the router computes the output port of the packet. This is followed by the *virtual-channel allocation* (VA), during which a VC is allocated for the head flit. With the known port and VC the stream competes for output physical channel in the *switch allocation* (SA) stage. When the stream is selected, the data is transferred on the output physical channel in the *switch traversal* stage. Depending on the implementation, some stages can happen in parallel. For example, if the RC stage computes the output port for the next router as the current port is already encoded in the header, the RC stage can happen in parallel with VA. Additionally, if fixed VC assignments are used, the VA stage can be omitted, such that RC and SA will be in parallel. Next to this typical router pipeline stages, an additional *buffer write* (BW) and *link traversal* (LT) phase might respectively be at the beginning and end, inducing additional delay.

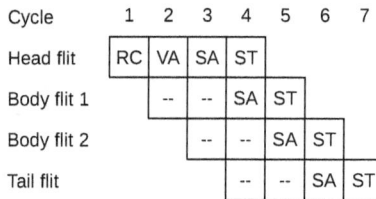

Cycle	1	2	3	4	5	6	7
Head flit	RC	VA	SA	ST			
Body flit 1		--	--	SA	ST		
Body flit 2			--	--	SA	ST	
Tail flit				--	--	SA	ST

Figure 2.6: *Typical pipeline stages (for a four flit packet).*

2.2 Selected NoC Architectures

There exist several implementations of NoC architectures. This section presents a brief description of some selected NoC architectures in alphabetical order followed by an enumeration of their most relevant architectural features in Table 2.2.

Aelite

Aelite is a modification of the Æthereal network architecture [90]. In contrast to Æthereal, Aelite only implements guaranteed service and skips the dedicated best-effort part. Additionally, it uses mesochronous links, avoiding the need for a global clock synchronization between the routers.

Arteris FlexNoC

The Arteris company provides a commercial NoC architecture for SoCs. The architecture is based on the *Danube NoC IP* library and most design parameters, such as topology, routing algorithm, flow control, and number of input and output ports, are user defined [23; 135]. The Danube library provides three major building blocks: *Network Interface Units* (NIU) for connecting IP blocks to the network, *Packet Transport Units* (PTU) constituting the network devices, and *physical links*. Based on this, Arteris provides a design flow and tools to create a customized NoC architecture.

Æthereal

Æthereal is a virtual-circuit switching network with time division multiplexing (TDM) and quality of service [81]. It uses two dedicated router parts, one for guaranteed service (GS) and one for best-effort (BE). In Æthereal the routers are time synchronized to allow the use of time slots, which form the most important mechanism of Æthereal. The time slots define which block of data can be forwarded through particular output of a router and for how long it is forwarded. Thus, time slots form schedules repeated periodically for the whole network, enabling a TDM based network scheduling. In such a way, the time slots avoid interference between different streams and provide a predictable behaviour of the communication leading to a *contention-free routing*. Best-effort traffic is allowed to use all unreserved or unused time slots. For BE traffic, wormhole switching and source routing are used with a credit based flow control.

CHAIN

The CHAIN network (CHip Area INterconnect) is an asynchronous (i.e. clock-less) network [25]. It uses priority based arbitration to offer bandwidth and latency guarantees [73; 74]. The arbitration is performed only during setting up the route, where CHAIN uses a source routing scheme. After setting up the route, the transfer relies on an asynchronous handshake backpressure scheme. CHAIN is based on cricuit-switching, hence after setting up a route, there is no sharing of the link among different traffic streams.

DyAD

The DyAD NoC is an 2D mesh topology without virtual channels that supports distributed adaptive routing and wormhole switching [102]. It applies deterministic or adaptive routing based on the state of the system. If there is no congestion, deterministic XY routing is used. When the network becomes congested, the routers switch to a different mode and apply an adaptive odd-even routing scheme. For this, each router is applied with a congestion flag, which is sent to neighbouring routers when the router is congested. If a router receives such a flag, it switches to the adaptive mode.

IDAMC

The IDAMC is a highly scalable NoC based architecture [128; 216] offering a general purpose system with support for applications of different (safety) criticality. It supports several models of SoC architectures: bus, crossbar, and full NoC. The network nodes are connected through NIs to the NoC. The NI are used as an abstraction layer and support address translation, monitoring and rate limiting. In general a NoC router has up to eight ports. Thus, it supports implementing different topologies. For small systems (four tiles or less) the router can be degraded to a crossbar, since the routing mechanisms are not needed [128; 216].

The NoC supports virtual channels and guaranteed service on pre-reserved connections. The QoS support is based on the "Back Suction" scheme [58]. For the routing of the data, source routing and wormhole switching are used. The routing is based on a flexible run length encoding of directions (in a mesh network) with up to three turns (i.e. four different directions). For the physical transport of the flits, the IDAMC uses serialisation and thus sends each flit as multiple phits over the network links. That way the link width can be decreased. The data and header information is protected using by a CRC [184].

iNoC

The invasive Network-on-Chip (iNoC) is a scalable, autonomous, self-optimizing and QoS-aware, interconnect architecture [93; 96; 97]. It implements a 2D mesh topology, but also supports other regular topologies, like 3D mesh or torus. For routing, iNoC supports two distributed routing algorithms: XY and adaptive odd-even-turn routing. QoS can be achieved using a weighted round-robin arbitration and end-to-end connections. Once an end-to-end connection is established, hard throughput guarantees can be given [93]. The iNoC uses wormhole switching and induces a static protocol overhead of 4–6 bit additionally to the destination address in the header flit and two bits in the tail flit.

Mango

The Mango (Message-passing Asynchronous Network-on-Chip providing Guaranteed services through OCP interfaces) NoC implements an asynchronous network providing a globally asynchronous locally synchronous system (GALS) [37]. Mango utilizes virtual channels with dedicated physical buffers and a router with dedicated parts for best-effort and guaranteed service traffic. The routers are organized in a 2D mesh, having five input ports. Latency and throughput are controlled with the usage of special arbitration modules for each network hop. Mango supports a fair or priority based arbitration scheme between different VCs. The configuration of all settings is done during the boot-up phase of the system. The transmission time in the network mainly depends on the number of VCs sharing a particular connection and the selected arbitration policy.

MPPA

The MPPA2-256 (Bostan) is a commercial architecture that is based on a NoC [42; 65; 66; 67; 190]. The MPPA architecture (Multi-Purpose-Purpose Processing Array) integrates 256 processing engine (PE) cores and 32 resource management (RM) cores on a single chip. These cores are divided into 16 compute clusters and four I/O subsystems. The MPPA uses two distinct physical network layers: a control NoC (C-NoC) and high-bandwidth data NoC (D-NOC). Both NoCs support full duplex links, wormhole switching, and output buffering without virtual channels. The routers have five ports and implement a 2D torus topology connecting the clusters and the I/O subsystems. The MPPA routers implement a round-robin arbitration and nodes offer rate limiters with a fine-grained traffic control. The flow regulation is adjusted for network calculus [134], and allows defining a set of linear constraints on the link bandwidths and on the router FIFO queue sizes.

Nostrum

The Nostrum mesh architecture is a virtual-circuit switching network with support for quality of service [150]. Nostrum is utilizes a time-division multiplexing (TDM) approach to provide guaranteed service. It implements *Temporarily Disjointed Networks* (TDNs) for temporal isolation between different traffic types (BE and GS) as well as between different virtual channels. In the VCs, *Looped Containers* (LCs) are used to provide guaranteed throughput [149].

The TDNs are based on "colouring" of outputs and inputs of routers in such a way that packets will move during time slots simultaneously using only their own colour assigned to a specific VC. This enables to ensure that no other packets will be placed in a particular buffer, allowing time multiplexing. LCs, on the other hand, are containers travelling from the source to the destination and back (in a loop) over a given VC filled with data or empty as place holders. The number of reserved containers for a VC corresponds to the throughput guarantee of this VC. LCs that belong to different circuits are allowed to share the same path. As the architecture is based on TDM, Nostrum does not require additional flow control. The network configuration is designed statically offline and initiated during system start-up, but allows some online modifications (e.g. the route is decided at design time but the bandwidth can vary at run-time).

Proteo

The Proteo architecture provides a parametrizable on-chip communication architecture [20; 189]. It targets hierarchical on-chip interconnects and thus provides, additional to the usual network components, bridges to interconnect sub-networks. Proteo supports different topologies, such as buses or star, but the main network is based on a bidirectional ring. It uses a destination-tag routing policy, where the destination address is stored in the packet header.

QNoC

The Quality of service NoC (QNoC) is a generic architecture for providing different levels of quality of service (QoS) at the communication layer [38; 39; 68]. QNoC is based on a 2D mesh topology and uses wormhole switching and XY source routing. The design assumes a reliable (physical) architecture and thus does not provide support for error correction. For quality of service, QNoC uses a priority based round robin scheduling on flit level. It distinguishes between four priorities, named service levels (SLs): *signal-*

ing (inter-tile control signals), *real-time* (hard delay constraints), *read/write* (short data access), and *block transfer* (large data bursts). The priority is assigned in the order as listed, with signaling the highest and block transfer the lowest priority.

RAW

The RAW processor uses four homogeneous 2D mesh networks to connect 16 tiles [210; 211]. In each tile, four routers are used, one for each network. Two routers provide a static network for single-word packets with in-order delivery following a route specified at design-time, flow-control, and reliable communication. The static network is targeted for instructions. The other two routers provide a dynamic network for interrupts and user-level messaging. It uses wormhole switching and a dimension ordered routing scheme.

SoCBUS

SoCBUS [234; 235; 236] uses *Packet Connected Circuits*, which is based on circuit-switching. Thus, SoCBUS operates on point-to-point connections owning exclusively all resources on a path. The connections are established by special configuration packets routed through the network, which reserve the resources. If a path is successfully reserved, a positive acknowledge is sent back and the data transfer starts. At the end of the transmission, a cancel message releases the reserved path. This results in a 4-phase transaction handling:

1. configuration packet from sender to sink, locking resources
2. acknowledge packet from sink to sender
3. data transfer from sender to sink
4. final acknowledge packet from sink to sender, cancelling the circuit

SoCBUS can instantiate any topology, but most examples use a 2D mesh topology. Here routers are connected with up to four ports to neighbouring routers and a wrapper is used to connect processing cores (via a fifth port) to a router.

SPIN

The Scalable Programmable Integrated Network-on-chip (SPIN NoC) implements a fat-tree topology with two one-way 32 bit data paths [18; 88]. The network uses wormhole switching and credit-based flow control. The flow control is on end-to-end basis between source and destination, where the destination has dedicated feedback wires to notify the sender of accepted data. SPIN uses an adaptive routing, such that routes can select any of the redundant paths to a destination to reduce traffic hot-spots. The routers use small (4 word) input buffers (to hide delay of control logic and link latency) and have two shared output buffers (18 word each) for blocked packets. The use of such shared buffers (compared to bigger input buffers) allows to reduce the overhead (needed buffer space) for networks/use cases where contention occurs rarely. The topology offers a non-blocking network and therefore no dedicated support for guaranteed service is provided (i.e. it uses round robin arbitration).

STNoC

STNoC is a commercial NoC architecture by STMicroelectronics [205]. It is based on the *spidergon* network topology and uses wormhole switching, output queuing, and deterministic source routing. Quality of service is achieved by using a fair bandwidth allocation scheme that enables to provide latency and throughput guarantees.

Tilera iMesh

The iMesh is a commercial NoC architecture by Tilera used in the Tile64 and TilePro64 [214; 233]. The iMesh is composed of multiple two-dimensional mesh networks. The Tile64 uses five and the TilePor64 uses six of these networks. In the basic design the links are 32 bit wide. Each mesh network handles a specific traffic type. These are: user-level messaging (UDN), I/O traffic (IDN), memory traffic (MDN), intertile traffic (TDN), caches coherence trafic (CDN), and compiler scheduled traffic (STN). The CDN network is only present in the TilePro64. The networks are classified in dynamic and static networks. The dynamic networks (UDN, IDN, MDN, TDN, CDN) use a distributed dimension ordered routing with the destination address encoded in the XY coordinates in the packet header. They use wormhole switching with a credit based flow control and without virtual channels. The routers implement a single stage pipeline during the straight portions of the routes, while there is an additional route calculation stage when turning. The static network (STN) uses circuit switching allowing high throughput streaming traffic. For this, a setup packet first reserves a specific route, the subsequent messages then follow this route.

XGFT

The XGFT (eXtended Generalized Fat Tree) NoC uses a fat tree topology and wormhole routing [109]. It supports *pipelined circuit switching*, which is an extension of the traditional wormhole mechanism. In this scheme, the header flit of a packet can be routed one step backwards when blocking occurs to take a different path using adaptive turn around routing. Additionally, XGFT is able to handle faults and to reconfigure the routers to provide some fault-tolerance. When a fault is detected, the routing is changed to a deterministic source routing, such that the packets can be routed around faulty NoC areas.

×pipes

Xpipes is a highly customizable architecture using wormhole switching and source routing [33; 55]. Due to the deterministic source routing, the routers can be kept simple and are similar to traditional virtual channel routers [54; 69]. The topology and hence also the number of input and output ports as well as number of virtual channels are design parameters. Various topologies, like mesh, torus, hypercube, clos, and butterfly, as well as routing algorithms, like dimension-ordered, minimum-path, traffic splitting across minimum-path, and traffic splitting across all paths, have successfully been evaluated using Xpipes. Xpipes supports a go-back-N retransmission strategy for link-level error control, where an error is indicated by a CRC module running concurrently.

Summary of SoA NoC Architectures

Table 2.2 presents an overview of the most important parameters of selected NoC architectures.

Table 2.2: Overview of Selected SoA NoC Architectures.

NoC	Topology	Routing	Switching/ Flow Control	Flit Size	Buffering	QoS
Aelite	mesh (any)	contention-free routing/ pipelined time-division-multiplexed circuit switching	no link level flow control	3 words	input (single word)	pipelined TDM/offline scheduling of flits
Arteris FlexNoC	custom	n/a	wormhole or S&F	custom	custom	Priority based
Æthereal (2002)	mesh (any)	BE: source; GS: contention-free routing/ pipelined time-division-multiplexed circuit switching	BE: wormhole, credit based; GS: TDM based	32 bit + 2 bit control	input queue	BE+GS router. GS: pipelined TDM/offline scheduling of flits
CHAIN	custom	source	circuit switching	—	—	priority based
DyAD	mesh	deterministic (XY)/ adaptive	wormhole	—	input	—
IDAMC	mesh (any with 8 port router)	source	wormhole/ credit based	140 bit (35 bit link width)	input	GT VCs

Table 2.2: Overview of Selected SoA NoC Architectures.

NoC	Topology	Routing	Switching/ Flow Control	Flit Size	Buffering	QoS
iNoC	mesh (other regular)	distributed XY and adaptive off-even-turn	wormhole	—	input	VCs with weighted-round-robin & E2E connections
Mango	mesh	deterministic	wormhole/ lock-based or credits	—	output	BE+GS router; GS: E2E reservation of VCs
MPPA	2D Torus	source	wormhole/credit based	32 bit	output	round-robin + source rate limiting
Nostrum (2001)	2D mesh	hot potato	S&F	128 bit data + 10 bit control	—	TDM based VC reservation
Proteo (2002)	bi-dir. ring	(distributed) destination-tag	circuit	variable	input and output	—
QNoC (2003)	2D mesh	XY	wormhole	16 bit data + 10 bit control	Input queue + single position output queue	GT virtual channels (four different traffic classes)
RAW	four meshes	design-time route/ dimension ordered	design-time/ wormhole	—	input / output	—
SoCBUS (2002)	2D mesh	XY adaptive	—	16 bit data + 3 bit control	single position input and output buffers	Circuit switching

Table 2.2: Overview of Selected SoA NoC Architectures.

NoC	Topology	Routing	Switching/ Flow Control	Flit Size	Buffering	QoS
SPIN (2000)	fat-tree	deterministic & adaptive	wormhole/ credit based	32 bit data + 4 bit control	input queues + 2 shared output queues	—
STNoC	spidergon	source	wormhole	—	output	fair bandwidth allocation
Tile64	2D Mesh	distributed XY	wormhole/ circuit switching	(32bit link)	input	—
XGFT	Fat-tree	adaptive	wormhole	36 bit	input/ output	—
×pipes (2002)	arbitrary	source static (street sign)	wormhole	32, 64 or 128 bit	virtual output queue	—

2.3 NoC Performance Verification

There exist many parameters influencing the performance (e.g. latency, throughput) of a NoC (cf. sections 2.1 and 2.2). For example, the designer needs to investigate the topologies, routing, switching, flow-control, buffer sizes, virtual channels, and link sizes (e.g. packet and flit serialisation). All these parameters influence the performance, predictability, robustness, and flexibility of the system. To understand the impact of the different parameters, their performance-cost tradeoffs, and the interdependencies between them, performance analysis and verification tools can be used. Applying these tools early in the design, enables to make design decisions an iterative process, where the feedback information can include functional (e.g. throughput, latency, jitter, reliability) and non-functional (e.g. network utilization, area, power consumption) measures. We can distinguish network performance verification methods in at least two categories: *simulation based* or *formal analysis*.

Simulation based approaches are frequently used when designing NoCs. They can be used for an evaluation of the general working of the design, to estimate the performance, and identify bottlenecks. Such an approach builds a network and traffic model and then simulates the functioning by loading the traffic into the network. The models can either be on an abstract level, only modelling the general behaviour (e.g. only routing and arbitration on the whole network level), or detailed by modelling all components in detail (e.g. internals of the switches and their interconnections). However, the level of detail can influence the speed of the simulation, where more details typically slow down the simulation [145]. A simulation can use synthetic or realistic traffic models or traces and can also include the interaction between the network and other resources to examine the performance-cost tradeoffs [173; 191; 192]. However, the applicability of simulation based approaches for verification with respect to safety is limited, as in a simulation there is typically no guarantee to observe the worst-case behaviour.

Formal analysis approaches, on the other hand, use mathematical models for the network and traffic to formally derive the behaviour of the system, including the best, worst and average case. Examples for analysis approaches are the schedulability analysis (response time analysis) [198], queueing theory [86], network calculus [53], and compositional performance analysis [99]. While some formal analysis approaches allow to derive bounds on the worst-case behaviour (e.g. maximum latency of messages), they typically use simplifications or (over-) approximations when modelling a system. This

Table 2.3: Overview of NoC simulators.

Simulator	Framework	Topologies	Heterogeneity	Synchronous/ Asynchronous
AdapNoC [108]	C++	Mesh/Torus	No	Synchronous
BookSim [105]	C++	Mesh/Torus/ Tree	No	Synchronous
DARSIM [141]	C++	All	Lim. (#VCs)	Synchronous
ENoCS [229]	Java	Mesh/Torus	Yes	Synchronous
Garnet2.0 [19]	C++ (gem5)	All	Yes	Synchronous
HNOCS [32]	Omnet++	All	Yes	Both
Nigram [104]	SystemC	Mesh/Torus	No	Synchronous
Nostrum [144]	SystemC	Mesh/Torus/ Tree/Ring	No	Synchronous
Noxsim [48]	SystemC	Mesh	Yes	Synchronous
NS-2 [127]	C++, OTcl	All	Yes	Synchronous
ORION 2.0 [107]	C++	–	No	–
Sicosys [175]	C++	Mesh/Torus	No	Synchronous
SunFloor [196]	SystemC	Mesh	No	Synchronous
Wormsim [136; 163]	C++	Mesh/Torus	No	Synchronous

is needed to cope with situations that are to complex to be expressed (easily) under mathematical models. However, this also leads to pessimistic results that might not be reached in the implemented system. For example, a derived maximum latency might not be able to occur in the real system. Chapter 3 provides more details on analysis approaches for NoCs.

Simulation and analysis methods can also be combined, for example to validate against each other or to speed up simulation based performance verification [131]. Table 2.3 presents an overview on commonly used simulation frameworks.

In this work, we use the *HNOCS* library [32] for the *OMNeT++* framework [226; 227]. *OMNeT++* is an object-oriented, modular, discrete event network simulation framework based on C++, which has gained widespread popularity as a network simulation platform in the scientific community (more than 200 *OMNEST/OMNeT++* related publications per year) as well as among industrial partners [12]. *OMNeT++* offers an easy integration of different libraries (e.g. various traffic sources) and an integration of on-chip and off-chip traffic. There are also several extensions for real-time simulation, network emulation, database integration, SystemC integration, and several other functions.

To model the NoC architectures in this work, we used and extended the open-source *HNOCS* library (Heterogeneous Network-on Chip Simulator) [32]. *HNOCS* offers a tool for modelling of heterogeneous NoCs with variable link capacities and number of VCs per unidirectional port. The *HNOCS* simulation platform provides a modular, scalable, expandable, and fully parametrizable framework with support for heterogeneous NoCs. For supporting performance evaluation and verification, the libraries offer a high number of tools for statistical measurements at flit and packet level: end-to-end latencies, throughput, VC acquisition times, etc.

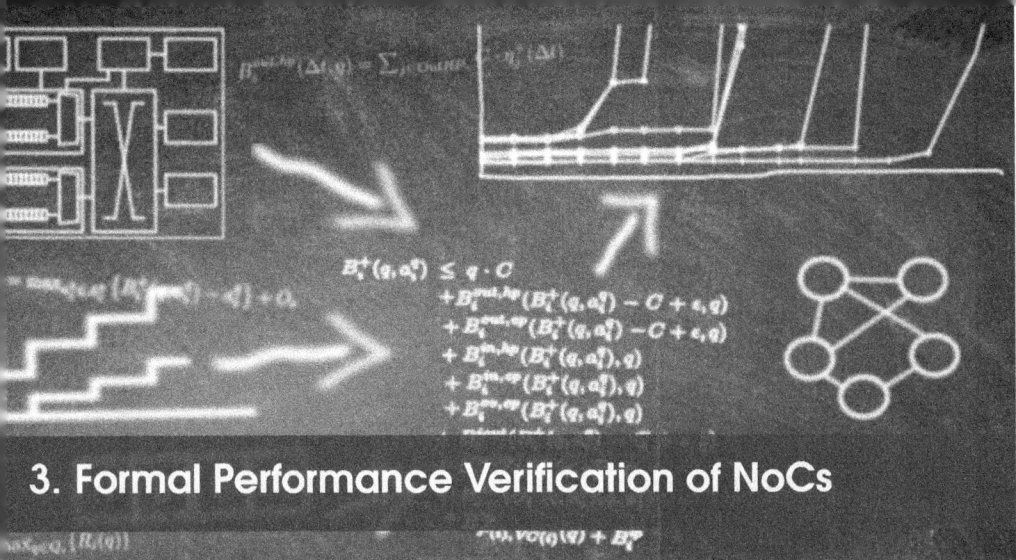

3. Formal Performance Verification of NoCs

In the domain of real-time systems, we need a performance evaluation or even verification of the system (cf. Chapter 1). Traditionally, such an evaluation of networks-on-chip (NoCs) is largely based on simulation. But for safety-critical or mixed-critical systems, a formal verification of the performance behaviour and hence formal guarantees for the worst-case behaviour of all real-time senders are needed. Thus, a simulation based approach is not suitable.

Most existing analysis approaches for NoCs are capable of providing such guarantees only under the assumption that the queues in the routers never overflow, i.e., that backpressure not occurs and a subsequent router or port can always accept incoming data. This leads to overly pessimistic guarantees or unfulfilled design requirements in many systems using commercially available NoCs where buffer space is limited. Therefore, this chapter presents an analysis methodology, providing formal timing guarantees for packet latencies also in a NoC where backpressure can occur. The analysis allows exploiting the behaviour of individual traffic streams to determine safe upper bounds on the latency of individual packets. The developed model can be used not only to obtain fast and accurate timing guarantees, but also to guide the NoC design process within an optimization loop. The accuracy of the analysis is evaluated experimentally through comparison with simulation results.

The chapter is partially based on the work published in [217; 220].

3.1 Introduction

In a typical design process, the application and platform models are designed independently [139; 140]. This independent development is done to cope with the design complexity of heterogeneous systems and to speed up the design process. The resulting application models are then mapped to the target platform. This mapping typically does not handle or solve all side effects and interferences that occur after integrating multiple applications to the NoC platform, as these effects only occur after the mapping is done. Hence, especially for safety-critical real-time systems, a performance and timing verification of the resulting system is needed. Simulations are not suitable for this, as they are slow and it cannot be verified that the worst-case behaviour was observed during the simulation, as required for safety-critical systems (cf. *freedom of interference* in ISO 26262 [7]). Hence, fast and accurate formal analysis approaches are needed. These need to adequately cover the actual platform architecture and provide fast and accurate results. The use of a fast analysis can then be combined with the mapping or even the design of the platform (e.g. layout, topology, and architecture). This enables, besides the individual verification of requirements, an iterative design and mapping process as sketched in Figure 3.1 [113; 139]. Such formal approach can be combined with the simulation-based approaches. For example, the analysis can be used in an iterative mapping to find a few possible mappings that satisfy all safety requirements in the worst-case. These few possibilities can then be investigated further in a simulation to obtain results on the average performance for selected use-cases and select the most promising candidate.

Most existing analysis approaches are capable of providing such formal verification only under certain simplifications. For example, these approaches assume an unlimited buffer space (i.e. that no backpressure occurs and a subsequent router or port can always accept incoming data), single flit buffers, no sharing of (virtual) channels between different streams, or no self interference (i.e. that a packet or even flit must have left the NoC before the succeeding packet or flit is released: $WCRT < D < T$). However, such simplifications lead to overly pessimistic guarantees or unfulfilled design requirements in many systems using commercially available NoCs where buffer space and the number of channels are limited. In this chapter we develop a backpressure aware analysis for a NoC with multiple virtual channels (VCs) and priorities. It supports a priority based arbitration between VCs with a different priority and a round-robin arbitration between VCs with the same priority as well

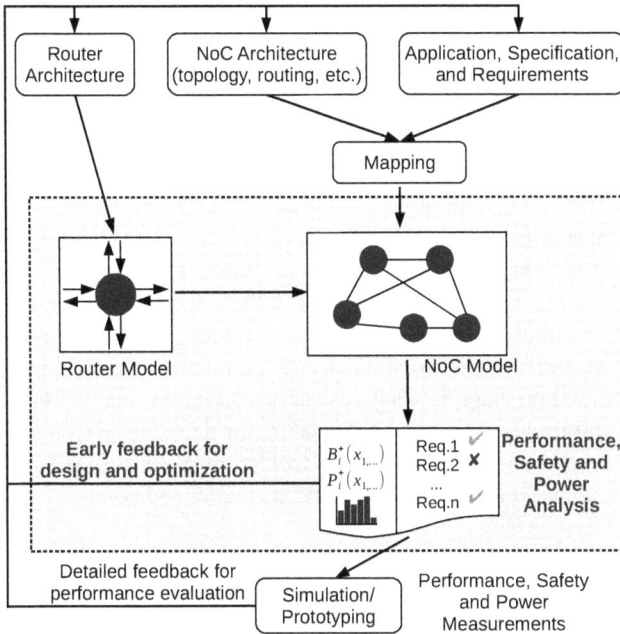

Figure 3.1: *Role of analysis during system design [113; 139; 164].*

as VC sharing between multiple traffic streams. The proposed network performance analysis approach provides the following performance metrics for each router and end-to-end:

- minimum accepted throughput;
- worst-case response times (WCRT) of single flits, single packets, and multiple packets;
- maximum buffer occupancies;
- output event models for tasks/streams (at each router).

And hence, the proposed analysis approach can be used for the verification of safety requirements, as well as for design and optimization purposes.

The remainder of this chapter is organized as follows. Section 3.2 provides an overview on related work. Section 3.3 briefly introduces the used analysis framework—the compositional performance analysis (CPA). Section 3.4 then presents the proposed analysis approach accounting for back-pressure. And finally, Section 3.5 presents an experimental evaluation of the approach, comparing the analysis results against simulation and another analysis.

3.2 Related Work

There exist various methodologies for the analysis of networks-on-chip including support for quality-of-service (QoS). For example, [27; 110; 115; 143; 198; 199; 200] provide techniques for timing analysis of priority-based wormhole switching NoCs. In [143] the authors formulate a contention tree that captures interference in the network. Similarly, Shi and Burns [198] define two different delay components: direct interference and indirect interference. Based on these, a worst-case network latency analysis is presented. However, all these schemes do not account for the effects of pipelining and parallel transmission of data. This is tackled by Kashif et al. [110], by refining the communication resource and its associated communication task model.

Still, the mentioned approaches assume global and unique priorities with unique virtual channel assignment for each priority. Such implementation policy typically results in high buffer cost and energy overhead. Hence, in many commercially available NoCs the foreseen number of (virtual) channels is usually lower than the number of tasks or priority levels. To address this problem, Shi and Burns [199; 200] allow shared priorities in a priority-based NoCs. While [199] requires that the deadlines are not larger than the period for all traffic-flows (and thus that a packet has left the NoC before the succeeding packet is injected), [200] removes this constraint.

In [56; 75; 76; 178; 180; 183] the authors present worst-case latency analysis approaches for networks using round-robin arbitration and with without special QoS support. For this, Ferrandiz et al. [75; 76] and Rahmati et al. [180] use a recursive calculus to obtain an upper bound on the traversal time of a packet. However, they do not take the individual behaviour of streams (e.g. inter-arrival time and periods of packets) into account, resulting in overly pessimistic results. This is addressed by Dasari et al. [56], where the authors extend the existing model by integrating the characteristics of the tasks that generate the packets. Another approach is presented by Rambo and Ernst [183], which uses a compositional performance analysis approach to analyse a NoC with *iSLIP* arbitration and shared virtual channels and hence allows non-symmetrical guarantees and to exploit behaviours of single streams (arbitrary event functions).

However, the approaches mentioned above assume that network nodes and routers are equipped with sufficient buffer space to prevent backpressure. For this, the network must be adopted to the particular application set, which is hard or even impossible, or the traffic injection-rate must be limited. As the rate limiting must be done according to the worst-case behaviour, this can lead to a decrease in system performance as the network cannot be fully utilized. To overcome these drawbacks, recent research focused on analysis techniques supporting backpressure [111; 177; 179].

Qian et al. [177] present an analysis for latency bounds in wormhole networks with finite sized buffers. The approach is based on network calculus [134] and computes global end-to-end service curves for each stream. Based on these performance parameters, as the maximum latency and minimum throughput, can be derived. This approach is close to ours. However, instead of computing global service curves, we follow the compositional approach of CPA [99; 100] and thus we can provide more fine-granular results, e.g., bounds for the behaviour at each router. This can help to more accurately account for interference and also to identify bottlenecks in the network. Kashif and Patel [111] present an extension of SLA [110] accounting for backpressure for priority based NoCs. And recently Nikolić et al. [162] present an analysis approach for priority-preemptive NoCs with per-traffic flow dedicated virtual channels that accounts for backpressure. However, the analyses assume unique priorities for each stream with individual virtual channels for each priority level and focuses on simple periodic activation models. On the contrary, our approach allows arbitrary activation models and sharing of the same (virtual) channel between different applications, as common for commercially available NoC architectures [17; 92; 233].

A slightly different approach is presented by Qian et al. [179]. Here the authors propose an average-case analysis for round-robin based NoCs. While the authors account for finite sized buffers (i.e. backpressure), the analysis does not provide safe upper bounds on the latency. The model only provides a fast approximation of the average latency as a fast alternative to simulation.

3.3 Compositional Performance Analysis (CPA)

3.3.1 Introduction

The proposed analysis is based on the *Compositional Performance Analysis* (CPA) framework [99; 100]. For modelling a system, CPA uses three main components: *resources*, *tasks*, and *event models*. Resources are used for modelling of processing or network nodes (e.g. CPUs, router ports). Tasks are mapped to resources and compete for service provided by the resources. The allocation of the service to the tasks depends on the selected scheduling policy.

To capture the dynamics of the system behaviour, task activations are abstracted using event models. These models define arrival functions $\eta^-(\Delta t)$ and $\eta^+(\Delta t)$, which provide lower and upper limits on the number of events (task activations) in any half-open time interval Δt. These have counterparts, the so called minimum and maximum distance functions $\delta^-(n)$ and $\delta^+(n)$. The minimum and maximum distance functions define the lower and upper limit on the time interval between the first and the last event of any sequence of n consecutive event occurrences. Both, η and δ, models can be straight-forwardly converted to each other [49; 99; 188; 194]. Figure 3.2 shows an example for both models. These models allow to capture all possible event arrival patterns within these limits. Thus, they cover all corner-cases and not only a single scenario as in case of a recorded memory trace.

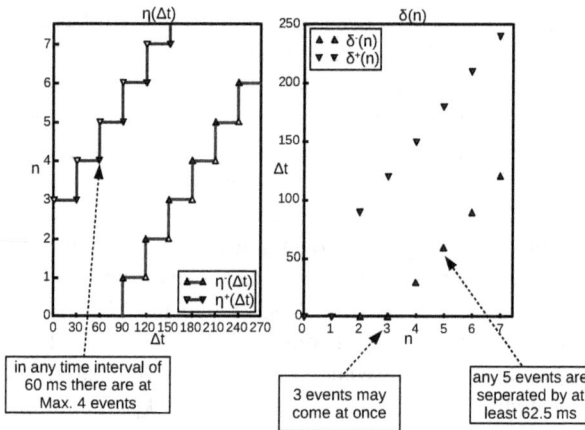

Figure 3.2: Event models covering all traces which stay between $\eta^+(\Delta t)/\delta^-(n)$ *and* $\eta^-(\Delta t)/\delta^+(n)$.

The execution of an application in a system is modelled using a directed graph. In such a graph, nodes denote tasks and edges symbolize dependencies between tasks. A task consumes temporal service of a resource, which can vary per activation between a best- and a worst-case execution time. The jitter (the difference between maximum and minimum response time) permits deriving new output event models of each task. Consequently, each element changes the event model, i.e., the output event model of a task on a particular resource becomes the input event model of its dependent task(s). Additionally, the CPA framework allows covering of functional dependencies between tasks such as task chains. For instance, activations of tasks that depend on inputs from multiple other tasks. This can be covered by joining the event models from several tasks using *AND* or *OR* functions [99]. CPA offers an iterative approach following the busy window method [215]. It minimizes and maximizes the response time of the currently analysed task by deriving the best- and worst-case blocking through all other interfering tasks, e.g., assuming the simultaneous activation of all interfering tasks and their maximum load (e.g. worst-case activation patterns). Figure 3.3 presents the approach. For this, CPA performs a local *busy window* analysis for each resource to compute worst-case timings and output event models for each task. The local resource-level analysis uses a *critical instant* scenario that assumes the worst-case arrival of all interfering tasks to obtain the maximum delay for the task under consideration. The output event models from the local analysis are then forwarded as input models for all dependent tasks and resources. With the new input models these tasks are then analysed again. The local analysis and propagation are iteratively applied until all output event models remain stable [99]. After each iteration, some output event models are updated, based on the analysis of the corresponding tasks. Tasks with updated activation event models are re-analysed. This potentially results in updated task output (and hence activation) event models in the next iteration and so forth. The initial event models for all tasks are derived from the external input event model of each task chain. In addition to the validation of the schedulability (i.e. all deadlines met) the analysis also yields upper bounds on the *worst-case response-time (WCRT)* of tasks and other timing properties such as the worst-case end-to-end path latency of a chain of tasks and the maximum backlog of activations.

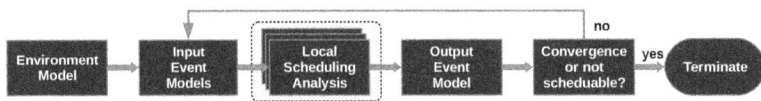

Figure 3.3: *Iterative analysis approach of CPA.*

3.3.2 CPA for NoCs

This section provides an overview on the *compositional performance analysis (CPA)* for NoCs. The CPA approach uses a similar composition and event models as *Real-Time Calculus (RTC)* [213], but differs in the used local analysis for the links and routers (cf. Section 3.3.1). For this, the *network-on-chip (NoC)* domain is translated to the processor resource model known from real-time scheduling [215]. This approach has been successfully applied to analyse on-chip networks [60; 63; 183; 198; 220]. The use of a CPA based framework enables to analyse heterogeneous systems using the same analysis and model. That is, the approach allows the routers to be different (e.g. to use different arbitration schemes, different serialisation of packets/flits, etc.) and to include the on-node/on-core behaviour to provide a system-wide end-to-end analysis.

The CPA framework uses a multicore processor model to represent the NoC [60; 183; 220]. In the model, *processing resources* represent the output ports of a router and *shared resources* with mutually exclusive access the input ports (or input virtual channels when each VC is directly connected to the crossbar). The exclusive access models the limitation of an input port (or VC) to send only one flit at a time to an output port (or output VC). A traffic stream is modelled as a *chain of tasks* mapped to the resources based on its path in the network. This is done through mapping the router ports to scheduled resources and the traffic streams to task chains that use a set of these resources based on their path in the network. Figure 3.4 shown an exemplary mapping of a router with four streams. In the example, streams 2 and 3, represented by tasks τ_2 and τ_3, share the input port and thus access the same shared resource InS. Stream 3 additionally shares the output port with stream 4.

In this model, the arrival of a flit is a task activation at the processing resource and the transmission of a flit at an output port is the execution of that task. Each task τ_i is assigned a best- and worst-case execution time C_i^- and C_i^+ for each task activation denoting the time needed to process a flit. The activations of a task can be triggered by an external source (network interface) or other tasks (routers).

In CPA, NoC analysis is performed iteratively using individual resource-level analysis steps to obtain the worst-case timing information. Figure 3.5 shows this approach. In the local analysis step, each router (i.e. each router output port) is analysed using the current input event models. Based on these models, the best- and worst-case response times and the jitter on each

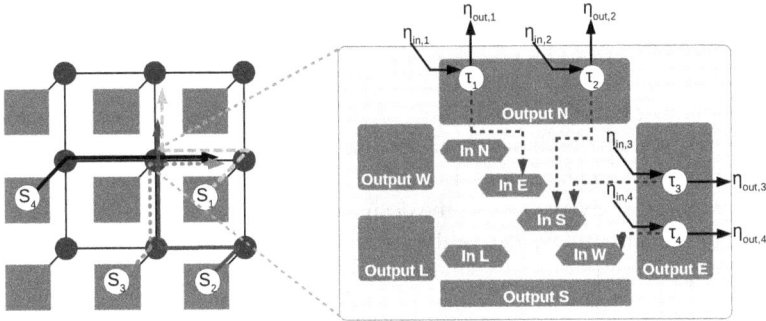

Figure 3.4: *Five-port router with four traffic streams as a multiprocessor with five processing resources (Output) and mutually exclusive shared resources (In).*

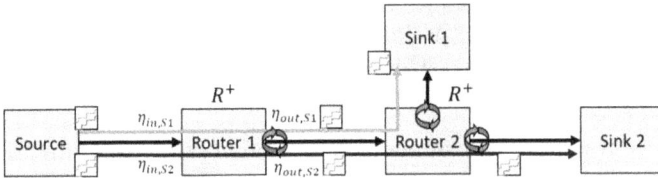

Figure 3.5: *Iterative CPA analysis of a NoC.*

router are calculated. These are then used to generate updated output event models for dependent tasks (on other routers). If the output or input models of a router have changed (compared to the previous iteration), the router is re-analysed with the new models. If the iterations remain stable, the response times for each task on each router are known. With these, an upper and lower bound on the end-to-end latency can be derived, e.g., by summing up the individual response times and accounting for additional timing parameters (cf. Section 3.4.4).

3.4 Backpressure Aware NoC Analysis

In this section we describe the proposed analysis of a network-on-chip with backpressure using CPA. We do this for a network-on-chip with multiple virtual channels (each VC having its own input buffers in a router), where we allow different and also the same priority for VCs, and also sharing of a VC between different streams. For the arbitration between different priorities we assume a priority based arbitration (packet or flit level) while requests with the same priority are handled using a round-robin scheme. While this is not a common design for existing NoC architectures, as this architecture is too complex, it allows to derive a general analysis approach that can cover most existing commercial designs, as e.g. [17; 92; 233]. That is, when applying the proposed analysis to a specific design certain terms will cease to exist.

For the analysis we derive the corresponding worst-case multiple activation processing time of a stream [63; 99; 220]. Based on this, we then derive metrics for a single router and for a complete network, such as the path latency.

> **Definition 3.4.1** The *worst-case multiple activation processing time* $B_i^+(q, a_i^q)$ of a stream i denotes the maximum time the resource (i.e. the router port) is busy processing q flits of stream i, given that all but the first flit arrive before their respective predecessor has been transferred and the q-th flit arrives at time instant a_i^q. The resource is considered busy if it processes a stream or if a backpressure signal to this resource is asserted.

To derive the worst-case multiple activation processing time, we first derive possible factors that influence the arbitration and hence the processing of a flit in the switch. Based on this, we then formally derive the actual influence these factors have on the processing time.

3.4.1 Influencing Factors

From Section 2.1.6 we know that a flit is only transferred from the input module of a switch to the output module if the following conditions are true: (1) to be able to request for the transmission, the flits must be at the front of the FIFO queue (i.e. all preceding flits inside the queue are already transferred); (2) the flit (i.e. its virtual channel) wins the arbitration at the input and output module; and (3) there must be sufficient space available at the next router (or the resource) to receive the flit (e.g. there must be available credits).

For the first condition, all flits that precede a stream inside the FIFO buffer must be transmitted. Figure 3.6 presents an example for this. When a flit of a stream i arrives at the network router, there might be already other flits in the FIFO queue waiting for transmission. These flits can be from the same or from different streams sharing the buffer. In the figure, for example, the flits of stream i are preceded by flits from a stream j that takes a different output port than stream i. Hence, stream i will have to wait until all these flits of stream j are transferred. Such a stream j may experience interference through other streams due to, for example, the arbitration at the output port. For the remainder of this work we will call such interference through other streams in the FIFO queue *FIFO blocking*.

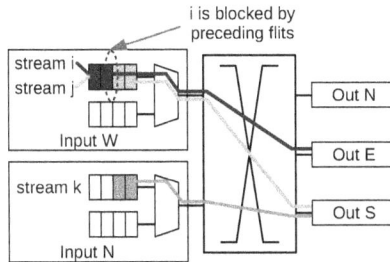

Figure 3.6: FIFO blocking

For the second condition, the arbitration policy at the input and output modules as well as the presence of other traffic streams influences when a stream i may win the arbitration. For the router described above, we can differentiate two different situations that influence the arbitration: traffic sharing the input port with a stream i and traffic sharing the output port with a stream i. Figure 3.7 shows an example for two streams sharing an input port (*Input W*). As the different virtual-channels are connected through a multiplexer to the crossbar, only a single VC can access the crossbar at a certain point in time. To handle the requests of multiple VCs, an arbitration is done between the VCs. With the definition of the router under analysis, a stream i may compete for access to the crossbar with other streams (i.e. other VCs) that have the same priority or a higher priority. Here, higher priority streams are always served before the stream i, while there is a round-robin arbitration between VCs with the same priority. If traffic on such VCs is present, it therefore might delay the processing of stream i. For the remainder of this work we will call such interference *input blocking*.

Besides the arbitration at the input module, there is an arbitration at the output module (cf. Section 2.1.6). Similar as for the input blocking, a stream i under consideration can be delayed by other streams at the output model. Figure 3.8 shows an example for this. As we already accounted for streams sharing the same input port, these interfering streams come from different input ports. With the definition of the router under analysis, a stream i may

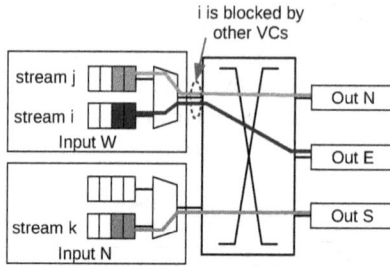

Figure 3.7: Input blocking

compete for access to the output port with other streams from other input ports that have the same priority or a higher priority. Similar as before, higher priority streams are always served before the stream i, while there is a round-robin arbitration between streams with the same priority. For the remainder of this work we will call such interference *output blocking*.

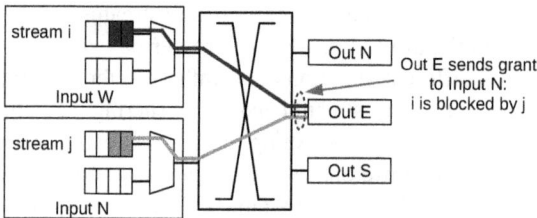

Figure 3.8: Output blocking

In addition to the input and output blocking, there is a special kind of blocking resulting from the independent arbitration at the input and output module. As discussed in Section 2.1.6, multiple input ports might request for the same output while each input port can send requests to multiple output ports. Hence, the arbiter of an output port can select a request of an input port but this input port might select a stream heading to another output port. Figure 3.9 shows an example for this. In the figure, streams i and j share the output port (*Out E*) and streams j and k the input port. If the shared output port now selects streams j but the northern input port (*Input N*) selects stream k, stream i is indirectly blocked by stream k. That is, a stream i can be delayed by a stream k that is not sharing the input or output port with the stream i. For the remainder of this work we will call such interference *indirect output blocking*.

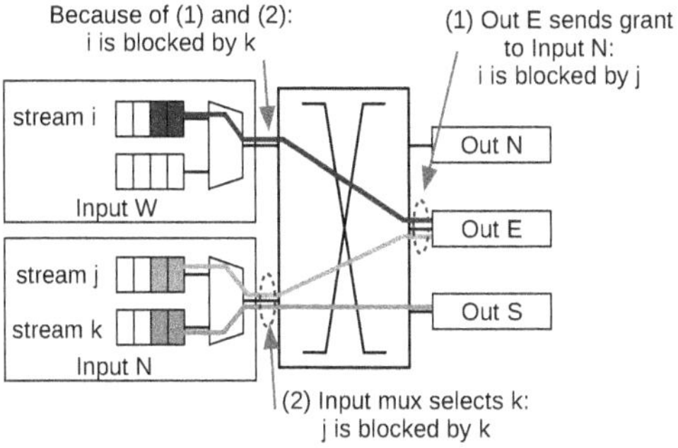

Figure 3.9: *Indirect output blocking example: stream* i *is blocked by stream* j *at the shared output port and stream* j *is blocked by stream* k *on the input port of stream* j.

For the third condition, there must be sufficient buffer space at the subsequent router (or the resource) to accept the arriving flits. As the processing of a flit can be delayed by many factors (see above), the buffer of a router might get full. In this case, the flow control will prevent the preceding router from sending additional flits. Hence, when looking at the flit processing of a router, the processing time is also influenced by the delay resulting from insufficient buffer space. Figure 3.10 shows an example for this. In the figure, a stream i at the current router cannot send its flits to the subsequent router as the input buffer is full. In such a situation, the buffer of the subsequent router may host flits of stream i or any other stream sharing this buffer. To be able to send a flit to the subsequent router, the current router has to wait until a flit at the subsequent router is processed (i.e. is fully transmitted). The processing at the subsequent router is again influenced by all factors discussed above, including the lack of free buffer space. For the remainder of this work we will call such interference *backpressure blocking*.

3.4.2 Sources of Blocking

With the known factors influencing the processing time of a flit, we now formally derive the worst-case multiple activation processing time. For this we first define various sets and auxiliary functions used throughout the remainder of this section. This is followed by the derivation of the

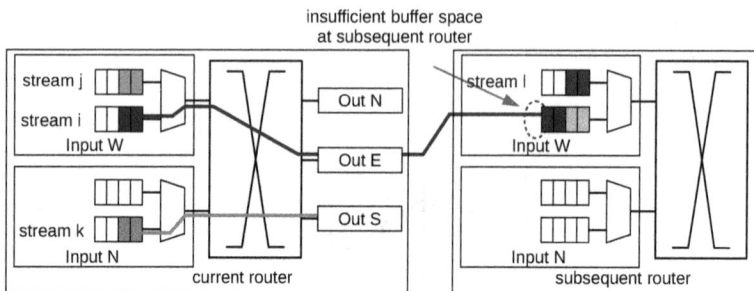

Figure 3.10: *Backpressure blocking: lack of free buffer space leads to a delay.*

individual blocking factors. For the different sets of streams, ports, and (virtual) channels we distinguish:

- Out_i^P: Set of all input ports accessing the same output port as stream i but not including the input port of stream i.
- $VC_{i,p}^{EP}$: Set of all VCs at port p that have the same priority as stream i without the VC of stream i.
- Out_i^{HP}: Set of all streams with a higher priority than stream i that share the same output with stream i but not the input port.
- In_i^{HP}: Set of all streams with a higher priority than stream i that share the same input port with stream i.
- $OutP_{i,p,c}^{EP}$: Set of all streams at input port p using VC c with the same priority as stream i that share the same output with stream i.
- $OutP_{i,p}^{HP}$: Set of all streams at input port p with a higher priority as stream i that share the same output with stream i.
- $InP_{i,c}^{EP}$: Set of all streams at the input port of stream i using VC c that have the same priority as stream i.
- Buf_i: Set of all streams sharing the (input) buffer with stream i.

Additionally, we use the following auxiliary functions

Definition 3.4.2 Let $\rho_i^+(\Delta t)$ be the maximum number of flits that can arrive in any time interval Δt at a stream i considering whole packets:

$$\rho_i^+(\Delta t) = \left\lceil \frac{\eta_i^+(\Delta t)}{n} \right\rceil \cdot n \qquad (3.1)$$

Definition 3.4.3 Let $\hat{\rho}(\Delta t, in, out, vc)$ be the maximum number of packets that can arrive in any time interval Δt at input port in on virtual channel c for output port out considering whole packets:

$$\hat{\rho}(\Delta t, in, out, vc) = \min \left\{ \left\lceil \frac{\Delta t}{n} \right\rceil, \sum_{j \in S(in, out, vc)} \rho_j^+(\Delta t) \right\}, \qquad (3.2)$$

where $S(in, out, c)$ denotes the set of all streams on input in using virtual channel vc and output out. In the equation, the first term of the min-function denotes the fact that there cannot arrive more packets than time has passed (assuming a flit transfer time of 1 cycle and the time to be in cycles). The second term then accumulates the maximum number of packets from each stream using input in, VC vc, and output out.

Definition 3.4.4 Let $\Theta(port, vc, k)$ denote the set of all possible mappings of k packets to streams using VC vc and port $port$. Then $\theta, \theta \in \Theta$ defines a specific mapping for k packets, such that θ denotes for each of the k packets the stream it belongs to and hence also the destined output port.

With these definitions, we can now derive the multiple activation processing time. To conservatively capture all possible worst-case scenarios, we break down the multiple activation processing time into a sum of different terms addressing all possible blocking factors. For a router as described above the processing time is influenced by:

- **Flit transfer time** C: the time to transfer a flit in a router excluding any kind of blocking. For the sake of simplicity and since it is the usual case in NoCs, we consider that all flits have the same constant transfer time of 1 cycle. However, other transfer times, e.g., when flits are transmitted in different numbers of phits, can be used.
- **Packet size** n: the length of a packet in flits. To improve readability, we assume all packets to have the same size, while the analysis can be straightforwardly extended to account for different packet sizes.
- **Buffer size** Q_b: the size of the buffer in the number of flits that can be stored. Without loss of generality, we assume all buffers to have the same size, while the analysis can be straightforwardly extended to account for different buffer sizes.
- **Output blocking** $B_i^{out,hp}$ and $B_i^{out,ep}$: the time streams from other inputs than i use the same output port.

- **Indirect Output blocking** B_i^{iout}: the amount of time that stream i is blocked by streams k from other inputs that do not share an input or output with i but directly block streams j sharing the output with i at their respective input.
- **Input blocking** $B_i^{in,hp}$ and $B_i^{in,ep}$: the time required to transmit flits from streams at the same input port but different VC than i.
- **FIFO blocking** B_i^{fifo}: the time required to transmit flits from other streams in the FIFO queue preceding the q-th flit of stream i.
- **Backpressure blocking** $B_{p,c}^{bp}$: the blocking resulting from lack of free buffer space at the downstream router or resource using port p and VC c.
- **LP blocking** B_i^{lp}: the time required to transmit the flits of a lower priority packet that already got selected.

With these we can define the worst-case multiple activation processing time in Theorem 3.4.1:

Theorem 3.4.1 — Multiple activation processing time. The worst-case multiple activation processing time $B_i^+(q,a_i^q)$ denotes the longest time required to transfer q flits of stream i on a router, given that all but the first flit arrive before their respective predecessor has been transferred and the q-th flit arrives at time instant a_i^q. It can be obtained by maximizing all blocking effects and summing them up:

$$
\begin{aligned}
B_i^+(q,a_i^q) \leq\; & q \cdot C + B_i^{out,hp}(B_i^+(q,a_i^q) - C + \varepsilon, q) \\
& + B_i^{out,ep}(B_i^+(q,a_i^q) - C + \varepsilon, q) \\
& + B_i^{in,hp}(B_i^+(q,a_i^q), q) \\
& + B_i^{in,ep}(B_i^+(q,a_i^q), q) \\
& + B_i^{iout}(B_i^+(q,a_i^q) - C + \varepsilon, q) \\
& + B_i^{fifo}(B_i^+(q,a_i^q), a_i^q) \\
& + B_{P(i),VC(i)}^{bp}(q) + B_i^{lp},
\end{aligned}
\tag{3.3}
$$

where $P(i)$ and $VC(i)$ denote the output port and virtual channel of stream i. The equation forms an integer fixed point problem, which is typical for busy-time based scheduling analysis. It can be resolved iteratively starting with $B_i^+(q,a_i^q) = q \cdot C$.

Proof. The theorem directly results from the definition of the NoC and its arbitration. The total multiple-activation processing time of q flits of a stream is bounded by the time required to transfer those flits ($q \cdot C$) plus the maximum time the stream may be blocked. Because flit-transfers cannot be preempted, the blocking only needs to be considered until the q-th flit begins transferring $(B_i^+(q, a_i^q) - C + \varepsilon)$. Adding ε is required because the η^+ function used inside the blocking terms to compute the number of interfering activations is defined as an open interval, i.e., $\eta^+(0) = 0$. ∎

Next, we derive upper bounds for the individual blocking factors from Equation 3.3.

Lemma 3.4.2 — Output Blocking with higher priority. The higher priority output blocking that a stream i experiences is bounded by:

$$B_i^{out,hp}(\Delta t, q) \leq \sum_{j \in Out_i^{HP}} C \cdot \eta_j^+(\Delta t). \tag{3.4}$$

Proof. Proven directly. Due to the priority based arbitration each stream with higher priority from other input ports that shares the output port with stream i and has flits waiting will be served before stream i if their output channel is ready. Hence, we need to account for the flit transmission time of all flits that arrive during the processing time of stream i. As higher priority streams are only served if their respective output port and VC is ready, we do not need to account for the backpressure blocking these streams may experience. ∎

Lemma 3.4.3 — Output Blocking with equal priority. The output blocking with equal priority that a stream i experiences is bounded by:

$$B_i^{out,ep}(\Delta t, q) \leq \sum_{p \in Out_i^P} \left(C \cdot \hat{\psi}_p(VC(i)) + B_{P(i),VC(i)}^{bp}(\hat{\psi}_p(VC(i))) \right)$$
$$+ \sum_{p \in Out_i^P} \sum_{vc \in VC_{i,p}^{EP} \setminus VC(i)} (C \cdot \hat{\psi}_p(vc)) \tag{3.5}$$

$$with \quad \hat{\psi}_p(vc) = \min\left\{ \left\lceil \frac{q}{n} \right\rceil \cdot n, n \cdot \hat{\rho}(\Delta t, p, OUT(i), vc) \right\},$$

where $IN(i)$, $VC(i)$, and $OUT(i)$ denote the input, virtual channel, and output of stream i.

Proof. Proven directly. With the definition of the arbitration and output blocking, there can only be output blocking of equal priority from two different types of streams: streams from other inputs using different VCs, and streams from other inputs using the same VC as stream i.

The first sum accounts for the blocking through streams with the same priority from other inputs using the same VC as i. Due to wormhole switching, once the scheduler grants access to an output port VC, no other input port can access this output port VC until it is released, i.e., the packet is fully transmitted. This is captured by $\hat{\rho}$, which considers that after a head flit arrives within the time interval Δt, the whole packet will be served before stream i. Additionally, due to the round-robin arbitration between streams with the same priority, each head flit belonging to stream i may only be blocked once by each other input port. This is addressed with the *min*-function, where $\lceil \frac{q}{n} \rceil$ is an upper bound on the number of head flits. Each of these head flits can be blocked at most for n flits from each other input port. Moreover, each of the interfering flits then will block stream i for the flit transfer time C and the backpressure blocking these flits experience.

The second sum/line accounts for the blocking from VCs with the same priority as stream i (but not using the same VC of i) from each input port (but not the input of i). Similar as for the first sum, these streams are served following a round-robin scheme. Hence, each of these streams can block stream i for each of its header flits. As these different VCs are only served when their respective output port and VC is ready, we do not need to account for backpressure.

As all of these streams share the set of higher priority output interferers, the interference of these is already accounted for when analysing stream i. Hence, we do not have to account for it again.

The sum of all possible blocking factors then derives the output blocking in Lemma 3.4.3. ∎

Lemma 3.4.4 — FIFO. The FIFO blocking that a stream i experiences is bounded by:

$$B_i^{fifo}(\Delta t, a_i^q) \leq m \cdot C + \max_{\theta \in \Theta(IN(i), VC(i), k)} \{A_\theta\}$$

$$+ \max_{\theta \in \Theta(IN(i), VC(i), 1)} \left\{ B_{P(\theta), VC(\theta)}^{bp}(m - k \cdot n) \right\} \tag{3.6}$$

$$with \quad m = \min \left\{ \left\lceil \frac{a_i^q}{n} \right\rceil \cdot n, \sum_{j \in Buf_i} \left\{ \rho_j^+(a_i^q) \right\} \right\}$$

$$k = \left\lfloor \frac{m}{n} \right\rfloor$$

$$A_\theta = \sum_{j \in \theta} \left\{ B_j^{out, ep}(\Delta t - C, n) + B_j^{iout}(\Delta t, n) + B_{P(j), VC(j)}^{bp}(n) \right.$$

$$+ \left. \sum_{k \in Out_j^{HP} \backslash Out_i^{HP}} C \cdot \eta_k^+(\Delta t) \right\},$$

where $IN(i)$ denotes the input of stream i; $VC(i)$ denotes the virtual channel of stream i; and k denotes the maximum number of whole packets (and hence head flits) of other streams.

Proof. Proven directly. The blocking caused by other streams in the same buffer consists of the flits that will be transmitted before stream i and the interference those flits observe. The first term accounts for the transmission of these flits. For this, only flits that arrived before the arrival of the q-th flit can be in the buffer. Thus, m provides the maximum number of flits of other packets that can be before the q-th flit of stream i in the buffer.

The interfering flits can be grouped into flits of whole packets as well as a packet partially transmitted at the front of the queue. The second term accounts for the blocking whole packets can experience. It considers for the k packets all possible mappings to streams and takes the maximum blocking. This blocking consists of output, indirect output, and backpressure blocking each packet may experience. For the output blocking with higher priority we only need to account for additional interference at other outputs (i.e. the ones not used by stream i), as for the same output it is already included when analysing the higher priority output blocking with of stream i. The third term accounts for the blocking of a partial packet. As the header of this packet has already been sent, we only need to account for backpressure blocking. ∎

Lemma 3.4.5 — Indirect Output. The indirect output blocking that a stream i may experience can be bounded by:

$$B_i^{iout}(\Delta t, q) \leq \sum_{p \in Out_i^P} (\psi_1(p) + \psi_2(p) + \psi_3(p)) \tag{3.7}$$

with

$$\psi_1(p) = \sum_{vc \in VC_{i,p}^{EP} \backslash VC(i)} \min \left\{ \left\lceil \frac{q}{n} \right\rceil \cdot n, \right.$$

$$\sum_{j \in OutP_i^{EP}, p, vc} \left\{ B_j^{in,ep}(\Delta t, \eta_j^+(\Delta t)) \right.$$

$$\left. \left. + B_j^{in,hp}(\Delta t, \eta_j^+(\Delta t)) \right\} \right\}$$

$$\psi_2(p) = \sum_{j \in OutP_{i,p,VC(i)}^{EP}} \left\{ B_j^{in,ep}(\Delta t, \rho_j^+(\Delta t)) + B_j^{in,hp}(\Delta t, \rho_j^+(\Delta t)) \right\}$$

$$\psi_3(p) = \sum_{j \in OutP_{i,p}^{HP}} \left\{ B_j^{in,ep}(\Delta t, \eta_j^+(\Delta t)) + B_j^{in,hp}(\Delta t, \eta_j^+(\Delta t)) \right\},$$

where $VC(i)$ denotes the virtual channel of stream i.

Proof. Proven directly. Indirect output blocking occurs when a stream j that shares the output port with stream i wins the arbitration at the output port but loses the arbitration at the input port to another stream. Hence, stream i may be blocked for the input blocking of such a stream j. This blocking can occur due to higher priority streams or same priority streams on the same or different VC than stream i. The first term accounts for all streams on different VCs on each other input port. Due to the round robin scheduling, stream i may be blocked by each of these streams for the number of the head flits, covered by the min-function. The second term accounts for streams using the same VC. Due to wormhole switching, we have to account for the blocking of the entire packet of the interfering streams. And the third term accounts for all higher priority streams from other input ports. As higher priority streams are always preferred at the arbiter, we need to account for all flits that can arrive during the considered time interval and their interference. ∎

Lemma 3.4.6 — Input HP Blocking. The input blocking caused by higher priority streams that a stream i may experience can be bounded by:

$$B_i^{in,hp}(\Delta t) \leq \sum_{j \in In_i^{HP}} \left\{ \eta_j^+(\Delta t) \cdot C \right\}. \tag{3.8}$$

Proof. Proven directly. Higher priority streams are served before stream i. Hence, for all arriving flits of higher priority streams, we need to account for their transmission. As these streams are only served (i.e. receive a grant from an output port) if the output channel is ready and no other input port receives the grant, we do not need to account for backpressure or output blocking of these streams. Additionally, as we already account for all higher priority streams when analysing stream i, we do not need to account for the higher priority streams of any stream j. ∎

Lemma 3.4.7 — Input EP Blocking. The input blocking caused by streams with the same priority that a stream i may experience can be bounded by:

$$B_i^{in,ep}(\Delta t) \leq \sum_{vc \in VC_{i,P(i)}^{EP}} \left\{ C \cdot \min \left\{ q, \sum_{j \in InP_{i,vc}^{EP}} \left\{ \eta_j^+(\Delta t) \right\} \right\} \right\}. \tag{3.9}$$

Proof. Proven directly. Due to the round robin arbitration each VC can send a flit for each flit of stream i. Hence, we need to account for their transmission time. As streams from other VCs are only selected, when their respective output port and VC are ready, we do not need to account for any blocking of these flits. For a "winner takes it all" arbitration (i.e. packet based arbitration between different VCs of the same priority) each other VC might send a full packet for each of the head flits of stream i. When using different packet sizes for the streams, we would have to select the worst (i.e. longest) candidates for the interfering streams. ∎

Lemma 3.4.8 — Low Priority Blocking. The low priority blocking a stream i may experience can be bounded by:

$$B^{lp} \leq \begin{cases} 0, & \text{if no lower priority (LP) streams exist,} \\ (n \cdot C) - \varepsilon, & \text{if LP exists and packet level preemption,} \\ C - \varepsilon, & \text{if LP exists and flit level preemption,} \end{cases} \quad (3.10)$$

where ε is the smallest time unit (e.g. a cycle) between the start of a flit transmission and the selection/start of a new flit.

Proof. Proven directly. When a flit of stream i arrives, the router might just have selected a low priority stream. Depending on the preemption policy, stream i then has to wait until a flit or a packet is fully transmitted. As the flit of stream i must have arrived at least ε time after the low priority flit was selected (otherwise the arbiter would have selected i), we can subtract ε. Additionally, as the low priority stream can only transmit when the output port is ready, there is no backpressure or additional blocking. For different packet sizes we would have to search for the worst candidate based on all low priority streams sharing the input or output with stream i. ∎

Definition 3.4.5 — BP. The backpressure blocking that a stream may experience on its path through output p and VC c in router k is given by:

$$B^{bp}_{p,c}(q) = \hat{B}^{+}_{p,c,k+1}(q), \quad (3.11)$$

where $\hat{B}^{+}_{p,c,k+1}$ is the worst-case waiting time for VC c at the downstream router $(k+1)$ until the q-th flit can be received. This accounts for the propagation of the new event model for backpressure in CPA.

Lemma 3.4.9 — Waiting. The worst-case waiting time of a port (i.e. buffer) p denotes the time until the port is ready to receive the q-th flit.

For a router it can be bounded by:

$$\hat{B}^+_{p,c}(q) \leq \begin{cases} q \cdot C + \max_{\theta \in \Theta(p,c,k)} \{A_\theta\}, & \text{if } b_p > Q_b \\ 0, & \text{otherwise} \end{cases} \tag{3.12}$$

$$\begin{aligned} \text{with} \quad A_\theta = \sum_{j \in \theta} \Big\{ &B^{out,hp}_i(\hat{B}^+_p(q) - C_i + \varepsilon, n) \\ &+ B^{out,ep}_i(\hat{B}^+_p(q) - C_i + \varepsilon, n) \\ &+ B^{iout}_i(\hat{B}^+_p(q) - C_i + \varepsilon, n) + B^{in,hp}_i(\hat{B}^+_p(q) - C_i + \varepsilon) \\ &+ B^{in,ep}_i(\hat{B}^+_p(q) - C_i + \varepsilon) + B^{bp}_{P(j),VC(j)}(n) \Big\}, \end{aligned}$$

where b_p denotes the worst-case backlog of the port and k is an upper bound on the number of packets q flits form ($k = \lceil \frac{q}{n} \rceil$). For router ports connected to a resource, the worst-case waiting time can directly be derived from the possible service (i.e. acceptance rate of flits) of the resource. This enables to use rate-limited resources that do not allow consuming a flit each cycle.

Proof. Proven directly. With the definition of the arbitration and waiting time, backpressure (i.e. waiting) can only occur, if the worst-case backlog of the port exceeds the buffer size (i.e. $b_p > Q_b$). If backpressure occurs, the port is conservatively assumed to be fully backlogged (i.e. buffer full). Hence, to receive q flits, the port must transmit q flits to any output. For these flits we must account for their transmission time ($q \cdot C$) and the worst-case interference they suffer. For this, the term $\max_{\theta \in \Theta(p,c,k)} \{A_\theta\}$ obtains the worst-case blocking from each possible mapping of flits to streams and output ports. ∎

With the lemmas 3.4.2 to 3.4.9, we have fully derived the inequality in Equation 4.1 for the local analysis step. For the iterative approach, we now define the derived metrics that are propagated between the routers and that can be used for performance and schedulability characterization.

3.4.3 Derived Metrics

With the worst-case multiple activation processing time, we can now derive the worst-case latency of a single router. The worst-case single hop latency

R_i^+ of a stream i denotes the maximum time between the arrival time a_i^q of the q-th activation (i.e. flit) and the processing of q activations $B_i^+(q,a_i^q)$ [61; 220]:

$$R_i^+ = \max_{q \in Q_i}\{R_i(q)\} \quad with \tag{3.13}$$

$$R_i(q) = \max_{a_i^q \in A_i^q}\{B_i^+(q,a_i^q) - a_i^q\} + O_r,$$

where O_r denotes the router's overhead, such as the time required by the router to determine and acquire the output port and virtual channel, and $R_i(q)$ is the worst-case response time of the q-th activation. This equation considers for each number of q activations within a busy-period all possible arriving times a_i^q of the q-th event. This is necessary, as a later arrival time might increase the interference in the FIFO queues. Additionally, a delayed arrival reduces the response time. However, the authors of [61; 183] proved that the number of possible candidates for a_i^q is finite. The authors showed, that it is sufficient to consider only candidates that coincide with activations of the interfering workload. Hence, we can define the set of candidates as:

$$A_i^q = \bigcup_{j \in I(i) \cup \hat{I}_i} \left\{ \delta_j^-(k) \mid \delta_i^-(q) \leq \delta_j^-(k) < B^+(q,a_i^q) \right\}_{k \geq 1}, \tag{3.14}$$

where $I(i)$ defines the set of all interfering streams for stream i using the same queue in the router and $\hat{I}_i$ represents a stream corresponding to the combined arrival curves of the interfering streams. This set again includes a fixed-point iteration as $B^+(q,a_i^q)$ is required to compute the set. For $\hat{I}_i$, the arrival curves of all interfering streams in the FIFO queue of stream i are combined and shaped such that only one flit can arrive per cycle. This additional factor (compared to the classic approach [183; 220]) is needed as the FIFO blocking (cf. Lemma 3.4.4) assumes that only one flit per cycle can arrive at the FIFO. Hence, for example, if there are two interfering streams that have a flit arriving at time instant 1, the set of $I(i)$ would obtain the arrival candidate of 1. Thus, the case where both streams are present in the FIFO queue is lost. The additional (virtual) stream $\hat{I}_i$ now leads to the arrival candidates of 1 and 2, covering the effects of multiple streams sharing the FIFO. In the classic approach this is tackled by assuming all interfering streams to have flits waiting in the FIFO when deriving the FIFO (e.g. input) blocking (i.e. that multiple flits can arrive in a cycle). However, that way the classic analysis assumes more interfering flits than physically possible.

Additionally, we need to consider all scenarios where an activation arrives within the busy-period of the previous activation when defining the number of activations q. Thus, we have to find all $q \geq 1$ that are smaller than the maximum number of events q_i^+ forming one busy-period [61]:

$$Q_i = \{1, 2, \ldots, q_i^+\} \quad with \tag{3.15}$$
$$q_i^+ = \min \left\{ q \in \mathbb{N}^+ \mid \delta_i^-(q+1) \geq \max_{a_i^q \in A_i^q} \{ B_i^+(q, a_i^q) \} \right\}.$$

With the worst-case waiting time, we can also derive the minimum accepted throughput $\hat{\beta}_p^-$ at a router port as:

$$\hat{\beta}_p^-(\Delta t) = \min \left\{ \lfloor \Delta t - C \rfloor, \max \left\{ m \mid \hat{B}_p^+(\max(0, m - Q_b)) < \Delta t \right\} \right\} \tag{3.16}$$
$$with \quad \Delta t \in \mathbb{N}^+.$$

The *max*-function selects the highest number of events that can be accepted during a time interval Δt based on the waiting time. For this, only events that can arrive before Δt can be accepted. As the first Q_b flits can be accepted immediately, we only have to account for the waiting time of $m - Q_b$ flits. Additionally, the port cannot accept more flits than can be physically be transmitted (e.g. one flit every C time units), covered by the first term of the *min*-function. The minimum accepted traffic for a sender then corresponds to the minimum accepted throughput at the first router port on the path of the sender. If this service is equal or higher the requested throughput in the long term, the stream is schedulable (w.r.t. to throughput requirements).

Based on the multiple activation processing time we can also derive the worst-case backlog in each buffer. The worst-case backlog of a port b_p of a router can be derived as:

$$b_p = \sum_{i \in Buf^p} \left\{ \max_{q \in Q_i} \left\{ \max_{a_i^q \in A_i^q} \{ \eta_i^+(B_i^+(q, a_i^q)) - q + 1 \} \right\} \right\}, \tag{3.17}$$

where Buf^p denotes the set of streams sharing the buffer p. This equation examines all numbers of activations q which arrive during the processing time of their predecessor. For each, $B_i^+(q, a_i^q)$ yields the completion time of the q-th activation. At this instance of time, at most $\eta_i^+(B_i^+(q, a_i^q))$ activations may have arrived since the start of the busy window, of which $q - 1$ have been processed. Hence, the difference yields the amount of backlogged activations, of which we have to determine the maximum for all $q \in Q_i$ providing the worst-case backlog of a stream. The sum of the worst-case backlogs of all streams sharing a buffer then returns the worst-case buffer backlog. However, due to backpressure (i.e. the flow control), this backlog only describes an analysis artefact used to estimate if backpressure can occur. The number of backlogged flits inside a router can never exceed the buffer size.

3.4.4 Analysis of Multiple Routers

With the analysis of a single router, we can now analyse a whole network using the compositional approach from Section 3.3.1. For this we iteratively perform a local analysis of the routers and propagate the event models (including backpressure) to neighbouring routers. Based on the local analysis, we can define the output events models for each stream that become the input event models of the subsequent routers according to [183] as:

$$\delta^-_{i,out}(q) = \max\left\{ (q-1)\cdot C, \delta^-_{i,in}(q) - (R^+_i - (C+O_r)) \right\} \tag{3.18}$$

where $(q-1)\cdot C$ denotes the best-case execution time, $\delta^-_{i,in}(q)$ the input event model, and $(R^+_i - (C+O_r))$ denotes the response time jitter (i.e. the difference between worst and best case execution times of a flit).

The worst-case waiting time $\hat{B}^+_p(q)$, as it is effectively propagated to preceding routers, is also an output event model—a new one. Hence, we extended the event (propagation) model of the CPA as sketched in Figure 3.11. Additionally, the worst-case waiting time influences the event model propagation of interfering streams and the blocking of the task under analysis. Hence, it takes part in other blocking factors. As backpressure depends on the downstream switch (or resource) the simple forward analysis (of CPA) becomes bi-directional. This can lead to cyclic dependencies that might lead to an infinite runtime of the analysis. Hence, this raises the question, whether there is a guarantee for the analysis to terminate.

The authors of [202; 204] showed that the CPA algorithm reaches the global fixed point under certain circumstances. First, the system-level analysis function shall execute all local component-level analyses at leas once. And secondly, all analysis functions, i.e. component-level analysis, computation and propagation of task termination event models, and the order in which local analyses are performed, must be order preserving. That is, for example, a component-level analysis (i.e. the analysis of a router port) must result in larger or equal results if any of its inputs increases. That is the case for the general CPA approach for NoCs as proved by Diemer [63]. Hence, we only need to prove that the backpressure blocking (i.e. the worst-case waiting time) is order preserving, i.e., monotonous increasing for increasing inputs (e.g. if $\forall x, y \in \mathbb{N}_0 : x \leq y \to \hat{B}_p^+(x) \leq \hat{B}_p^+(y)$ is true). With the definition of the worst-case waiting time (cf. Lemma 3.4.9) we can differentiate three cases: (1) there is no waiting as the buffer is not full ($b_p <= Q_b$); (2) there is always some waiting as the buffer is full ($b_p > Q_b$); (3) the buffer state (*full* or *not full*) changes during the analysis.

For the first case, the condition is always true as Equation 3.12 results in a constant value. For the second case, Equation 3.12 sums up other blocking factors. These factors are monotonous increasing functions [63]. And as the sum of monotonous increasing functions is a monotonous increasing function, the worst-case waiting time is increasing too. For the third case, we need to derive the possible state changes of the buffer. From the definition of the backlog (cf. Equation 3.17) and the event-models (η^+ [63; 99; 204]), we know that the backlog is increasing. Hence, the buffer state can only change from *not full* to *full* during the analysis and not vice versa. That is, during the analysis we might first have some iterations leading to the first case, followed by iterations leading to the second case. And from the definitions of the other blocking factors, we know, that the worst-case waiting time for the second case always results in values higher or equal zero and the condition is fulfilled. Thus, the worst-case waiting time is monotonous increasing (i.e. it never decreases).

Figure 3.11: *CPA model propagation extension.*

With the input models of all routers, we can limit the worst-case end-to-end latency $l_p^+(q)$ for transmitting q flits on a path p for each stream. It consists of the worst-case response time for each hop on the path p, the time to inject the q flits, and the packetization overhead. We can differentiate two different worst-case latencies: (1) when injecting q flits at any time (i.e. assuming there is already backlog); and (2) when injecting q flits assuming the network interface to be idle (i.e. no backlog when starting the transmission).

For injecting the q flits at any time (e.g. with backlog) the latency can be derived by:

$$l_p^+(q) \leq \max\left\{ \delta_{First(p)}^-(q), \hat{B}_{P(First(p)),VC(First(p))}^+ \left(q + b_{P(First(p))} - Q_b\right) \right\}$$
$$+ O_p + \sum_{j \in Tasks(p)} R_j^+, \tag{3.19}$$

where $First(p)$ defines the first task of the chain (i.e. network path); $Tasks(p)$ denotes the set of all tasks of path p (i.e. one per hop); O_p denotes the constant de/packetization overhead; R_j^+ denotes the worst-case single hop latency; $\delta_{First(p)}^-(q)$ denotes the time the sender needs to inject q flits (assuming no contention); Q_b is the size of the buffer; and $\hat{B}_1^+(q + b_i - Q_b)$ denotes the overhead induced by backpressure until the q-th flit and all backlogged flits of previous transmissions can be injected to the first router.

For the seconds case, when the first buffer is assumed to be empty, the latency can be derived by:

$$l_p^+(q) \leq \max\left\{ \delta_{First(p)}^-(q), \hat{B}_{P(First(p)),VC(First(p))}^+ \left(\max(0, q - Q_b)\right) \right\}$$
$$+ O_p + \sum_{j \in Tasks(p)} R_j^+. \tag{3.20}$$

In this case, we do not have to account for the waiting time of already backlogged flits but only for those flits that do not fit into the first buffer, covered by the max-function in $\hat{B}^+$.

Basically the equations compute the time interval required by a stream to inject q flits when the sender is fully backlogged or idle and then assume the last one of these to experience the worst-case blocking on all intermediate routers. Due to the in-order delivery of the network, all previous flits will have arrived at the destination before the last one. And the delay previous flits may observe is included as interference in the worst-case blocking of the last flit.

3.5 Evaluation of the Analysis Approach

In this section we evaluate and compare the analysis from Section 3.4 against simulation results. The results were obtained using the *OMNeT++* framework [226; 227] and the *HNOCS* library [32]. For the evaluation we use the simple system shown in Figure 3.12. This system is compact enough to be comprehensively displayed but shows all relevant effects of the analysis. It consists of four streams periodically injecting traffic from source Sx to destination Dx with a packet size of four flits. Each router induces a four cycle routing overhead (O_r) to the flits and has buffer space for two packets.

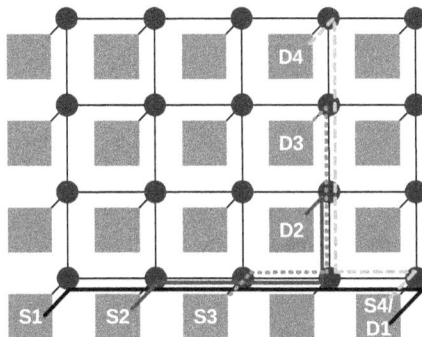

Figure 3.12: *Setup with four different streams, each sending from source Sx to destination Dx.*

In a first experiment we varied the requested bandwidth of each sender, where all senders request for the same bandwidth. Figure 3.13 shows for this, for each of the streams, the analysed and simulated worst-case flit end-to-end latency when using the same virtual channel for all traffic streams. The analysis for all streams for a single configuration took in average 908 ms. The results show that the analysis delivers conservative results for the experiment. For low bandwidths (up to 10 % per sender) the results from analysis are comparatively accurate w.r.t. simulation. For higher loads, the influence of backpressure and head-of-line blocking lead to a higher over-approximation for streams 1 and 2. For streams 3 and 4 the analytic results are much tighter, as these streams compete for a lower number of links and, thus, experience lower interference. Due to the over-approximation of blocking, the analysis reaches the saturation point earlier. This point is the bandwidth at which the backlog of a sender, and hence its latency, goes to infinity due to the NoC load.

For the same setup, Figure 3.14 compares the requested and obtained throughput per sender. As long as the analysis provides results for the latency (i.e. before network saturation), the analysed throughput is similar to the simulated. As soon as the saturation point (for a stream) is reached, the analysed throughput diverges from the simulation and remains on nearly

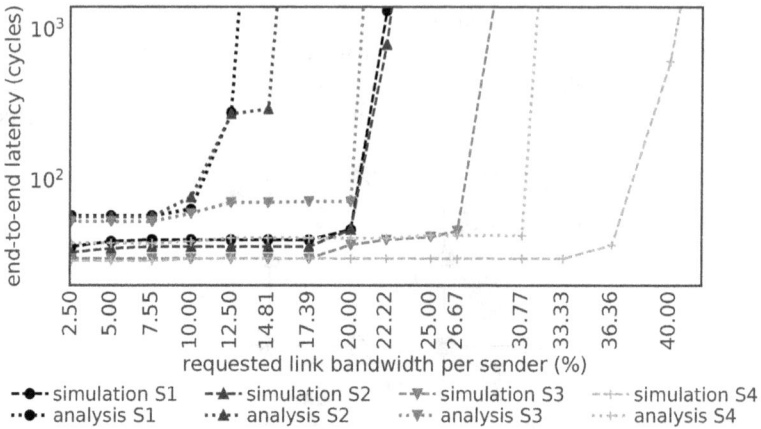

Figure 3.13: *Flit worst-case latency over requested bandwidth (per sender) with a buffer depth of eight flits, a packet size of four flits and all streams sharing a single VC.*

constant level. Along with the increasing load, the simulated throughputs start to decrease and to converge to the analysed values. This is especially visible for streams 1 and 2. This shows that for systems with a shared channel the complex blocking scenarios hinder tight latency bounds in analysis but nonetheless permit accurate bandwidth estimations. We can also observe, that the accumulated throughput of all senders is always below 100 %. This results from the effects of head-of-line blocking and backpressure, which render the full usage of available throughput impossible in most use cases.

In Figure 3.15 we vary the buffer depth and compare the simulated latency against the analysis from Section 3.4 and the basic *iSLIP* analysis of Rambo and Ernst [183] that assumes infinite buffers. For this experiment we used a packet size of four flits and requested load of 12.5 % for each sender. For small buffer sizes, our analysis delivers a high over-approximation of the latency. This is because for small buffers the head-of-line blocking and backpressure occur more likely and propagate faster through the system. Hence, the conservatism of the analysis and blocking propagation have a higher influence on the results. For larger buffer sizes, the results of the analysis become tighter, as backpressure is lower or even disappears, diminishing the negative effect of blocking propagation. Indeed, for buffer sizes greater than eight packets, our analysis delivers results as tight as the one

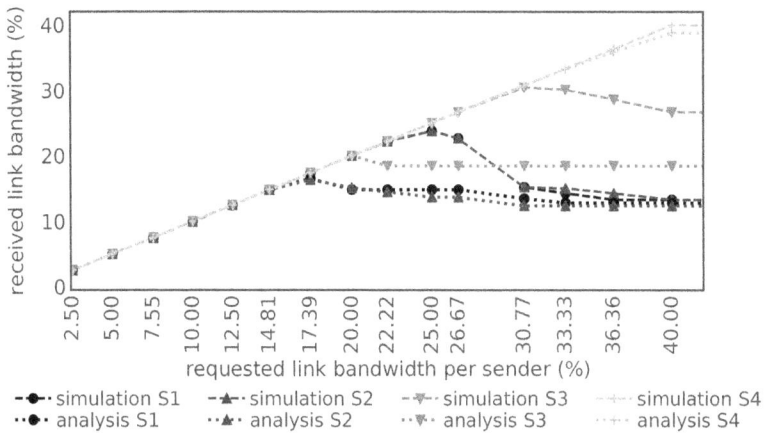

Figure 3.14: *Received versus requested bandwidth (per sender) for a buffer depth of eight flits and a packet size of four flits.*

of Rambo and Ernst [183] for the experiment. However, the results of Rambo and Ernst [183] are not proved to be safe for this setup as backpressure can occur.

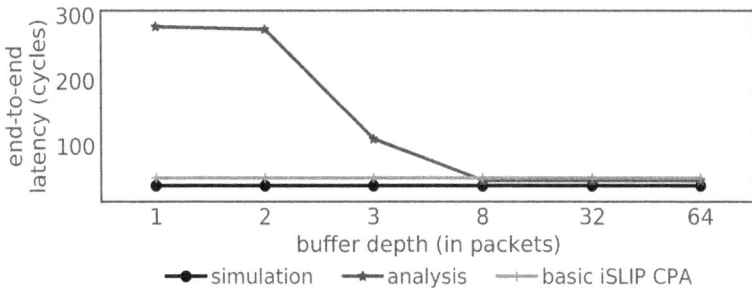

Figure 3.15: *Flit worst-case latency for stream S1 for different buffer depths and 12.5 % requested bandwidth per sender.*

For the same setup Figures 3.16 and 3.17 show the analysed and simulated worst-case flit end-to-end latency when using different VC assignments. For Figure 3.16 we assign stream S2 to a high priority VC while the other streams are sharing a single low priority VC. For Figure 3.17 we assign stream S2 to a high priority VC while the other streams are sharing the same priority level but use dedicated VCs (i.e. one VC for each stream). The results are similar as before and the analysis delivers conservative results for the experiments. Contrary to the results before, we do not reach the saturation for stream S2 (when only requesting up to 40 % of the link bandwidth). This is due to the high priority of S2, which avoids high interference from the other streams. The results also show, that an increased number of VCs can help to reduce the interference experienced by a stream in the simulation as well as in the analysis.

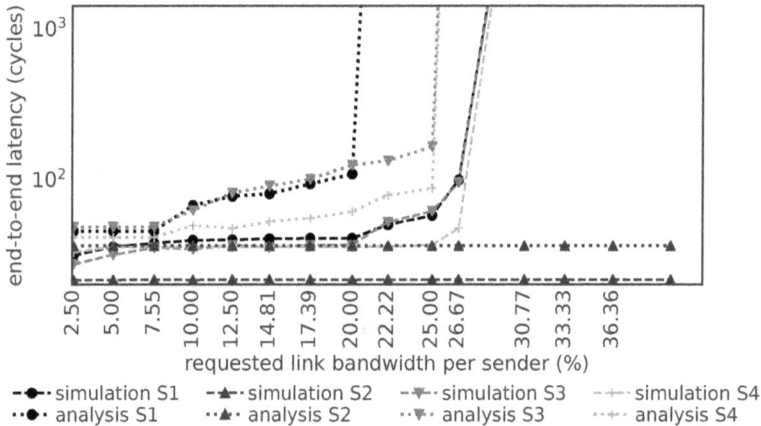

Figure 3.16: *Flit worst-case latency over requested bandwidth (per sender) with a buffer depth of eight flits and a packet size of four flits, where S2 uses a higher priority VC and the other streams share a VC.*

The evaluation shows that our analysis provides safe upper bounds on the worst-case flit latency for a NoC with backpressure. However, it also shows that backpressure can lead to overly pessimistic guarantees for a system. This has two reasons. First, the analysis is pessimistic when accounting for the interference. It does not consider correlations or the *pay bursts only once* [134] phenomenon, but applies an over-approximation. This leads to an analytical worst-case that can never happen in the real system. For instance, when sending a flit, the analysis assumes the worst-case backlog to occur during injection (i.e. backlogged sender) followed by the worst-case end-to-end latency for the flit transmission. However, this assumes the worst-case interference to happen twice: when building the backlog and when the flit flows through the NoC.

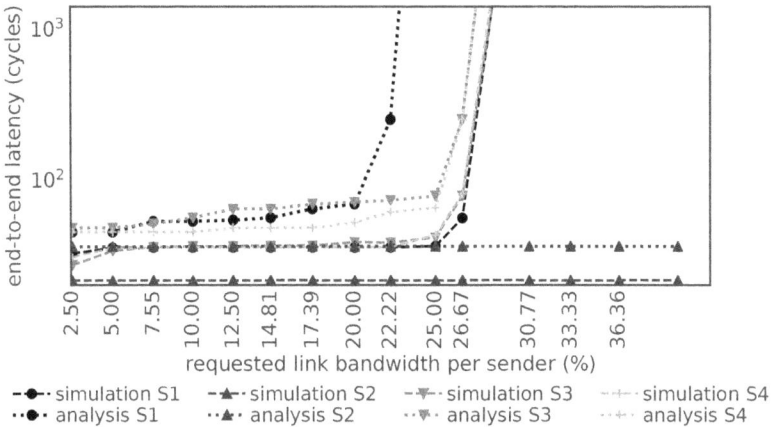

Figure 3.17: *Flit worst-case latency over requested bandwidth (per sender) with a buffer depth of eight flits and a packet size of four flits, where S2 uses a higher priority VC and the other streams use dedicated VCs with the same (low) priority.*

Second, backpressure constitutes a significant problem in systems with shared channels. Analyses typically introduce pessimism that will be increased when accounting for backpressure. The adverse effects of backpressure and blocking can also be seen in simulation (or real) systems, as, for example, shown in Figure 3.14, where the system could not be fully utilized. Hence, for real-time systems, the concurrent access to a shared channel between multiple real-time senders must be avoided or restricted, as, for example, with the use of rate limiting for the injection rates. This permits limiting the interference and improving the analysis results as shown in Figure 3.13.

3.6 Summary

This chapter provided an overview on the formal verification of the performance behaviour of safety-critical or mixed-critical NoCs. Most existing NoC analysis approaches are capable of providing formal performance guarantees only under the assumption that the queues in the routers never overflow, i.e., that backpressure not occurs. This leads to overly pessimistic guarantees or unfulfilled design requirements in many system using commercially available NoCs where buffer space is limited. To solve this problem, this chapter presented a new analysis methodology that provides formal timing guarantees for packet latencies also in a NoC where backpressure occurs. The analysis is based on the CPA framework. It exploits the behaviour of individual traffic streams to determine safe upper bounds on the latency of individual packets. The developed model can be used not only to obtain fast and accurate timing guarantees, but also to guide the NoC design process within an optimization loop. The accuracy of the analysis was evaluated experimentally through comparison with simulation results. Hence, we showed that the CPA framework can be applied for real-world systems where buffer space is limited, such as the Kalray MPPA-256 [42; 65] or Tile64 [214; 233].

However, the evaluation also demonstrated that backpressure and blocking propagation can lead to overly pessimistic analysis results, especially for systems with shared buffers. From this we can follow two aspects. First, the analysis approach needs further refinements to capture the system behaviour more accurately. That way the conservatism might be reduced, which possibly leads to better results (e.g. results that are closer to the real-world/observed behaviour). Secondly, systems with limited, shared buffer space need advanced quality of service (QoS) mechanisms that can optimize the buffer usage. With this, the conservatism of the analysis can be decreased and the adverse effects of blocking propagation (also for simulation and real systems) be reduced. Such mechanisms can be used to decrease the worst-case contention that must be assumed by the analysis and to reduce blocking propagation. Both factors can lead to better analysis results and even to an improved performance.

4. Quality of Service in NoCs

Networks-on-chip (NoCs) for future mixed-criticality systems must handle a growing variety of traffic requirements, ranging from safety-critical real-time traffic to bursty latency-sensitive best-effort traffic. This requires mechanisms to provide sufficient independence to enable an efficient design (cf. Chapter 1). Sufficient independence is typically achieved with quality of service (QoS) mechanisms implemented in or on-top of the network architecture. However, most existing solutions for performance isolation degrade the flexibility of the system or the performance of the non safety-critical senders. But this degradation is often unnecessary.

In this chapter we present run-time configurable QoS designs for a NoC with reduced adverse impacts on the performance of best-effort traffic. We show that the overhead implementing our approaches is affordable. And through experimental evaluations, we show that the approaches reduce the adverse effects through strict prioritization on best-effort applications.

The chapter is partially based on the work published in [125; 219; 221; 223].

4.1 Introduction

Multiprocessor systems on chip (MPSoCs) must frequently accommodate heterogeneous workloads with different timing and safety requirements, forming mixed-criticality multicore systems [28; 45]. The challenge in integrating mixed-critical applications comes from the multicore inherent architecture. Shared resources, such as the communication infrastructure or memories, couple the execution behaviour across cores. This impacts non-functional system properties like timing, which are of particular interest in safety-critical environments. Safety standards explicitly address this problem and require *sufficient independence* between functions of different criticalities (e.g. IEC 61508 [15], ISO 26262 [7]).

NoCs are foreseen as a promising interconnect solution for MPSoCs (cf. Section 1.1). However, in a NoC resources, such as output ports in routers, are typically shared among different functions and safety classes. Hence, applications of different safety levels will inevitably compete for resources in a NoC. One approach to solve this problem is to develop all functions to the highest relevant safety level. However, this leads to higher development costs and lower system utilization. Therefore, it is crucial to provide quality of services (QoS) mechanisms guaranteeing *sufficient independence* in a NoC. Figure 4.1 sketches the effect of QoS mechanisms. In the figure, traffic stream *B* is claiming too many resources, such that the other two streams (*A* and *C*) will not get sufficient service (top). When using QoS, stream *B* cannot negatively disturb the service of the others (bottom). However, as stream *B* is now not receiving all requested resources, it may be slowed down, compared to the first scenario.

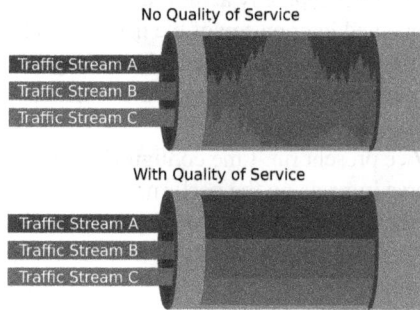

Figure 4.1: Exemplary quality of service (QoS) effect.

Most of today's NoC designs achieve the independence through static resource over-provisioning or static prioritization of safety critical traffic (cf. Section 2.2). This typically leads to a degradation of the performance, especially for non-safety critical applications. But for most safety-critical applications, it is sufficient to arrive shortly before or by their deadline and an early arrival provides no benefits [201]. On the other hand, a low latency for best-effort (BE) traffic can drastically increase the performance of BE applications [156; 217]. Hence, this performance degradation, when using static prioritization or static over provisioning, is often unnecessary. The goal of this chapter is to present new QoS mechanisms that have a reduced adverse impact on the system performance.

The remainder of this chapter is organized as follows. Section 4.2 provides an overview on related work on providing QoS in NoCs. Section 4.3 presents an efficient mechanism to provide guaranteed latency. And Section 4.4 describes the provisioning of guaranteed throughput. And finally, Section 4.5 briefly presents an approach reducing the hardware complexity and increasing flexibility through moving the contention control to the boundaries of the network (e.g. to the network interfaces).

4.2 Related Work

There exist various packet-switched networks-on-chip providing quality-of-service (QoS) for mixed criticality systems that can be categorized by how they enforce service guarantees (cf. Section 2.2). One group uses time-division multiple-access (TDMA) to limit the interference between applications [84; 149; 174]. These rely on a pre-allocation of timeslots in the network. Through assigning different safety classes to different time-slots, they provide strong isolation between safety classes. The Nostrum [149] architecture uses *looped containers* that are continuously routed through the network. Applications are statically mapped to certain containers to isolate them. Aetheral [84] defines schedule tables in each node and router. These are used to define a static schedule for the whole communication in the NoC. The DPSIN [152] network combines rate control and TDMA. For providing guaranteed service (GS), a virtual channel (VC) in every switch is reserved for GS traffic as a TDMA channel. In this TDMA channel, guaranteed service is obtained by allocating time slots to the streams.

The fixed allocation of timeslots often leads to an inflexibility and a high static overhead, as each packet might have to wait for its timeslot at multiple network nodes, even when other timeslots are empty. This inflexibility makes these architectures not suitable for high performance mixed-criticality systems. PhaseNoC [174] tackles this problem by defining scheduling in form of *waves* through the network. Through aligning the timeslots of all routers on a path, packets can travel with a reduced latency through the NoC. Another approach is the concept of channel-trees [89]. In this scheme, time slots are shared between a selected set of applications, to utilize otherwise unused timeslots. However, as these approaches rely on static timing schedules and slot assignments, they still introduce inflexibility.

To relax the conservatism and timing overhead introduced by TDM, more flexible approaches were proposed. These are based on the idea of combining interface-based design and system analysis [29; 231]. Components and applications provide well-defined interfaces and therefore introduce a bounded interference to the system. With this, more dynamic and work-conserving QoS mechanisms can be constructed based on rate controlling and dynamic scheduling [35; 39; 44; 93; 103]. Bolotin et al. [39] propose the QNoC architecture. It uses four traffic classes and a fixed priority scheme in the switches to arbitrate between packets of the different classes. To isolate critical traffic from best-effort traffic, the critical traffic has a higher priority. Hence, it blocks best-effort packets whenever these compete for network resources. In the Mango NoC [35] switches consist of two parts, a best-effort (BE) and a guaranteed service (GS) switch, and implement virtual channels. The GS streams are prioritized over BE streams and a fair-sharing arbitration is used between multiple GS streams. The latency of a message is bounded and mainly depends on the number of VCs sharing a particular connection and the selected arbitration policy. The Kilo-NoC [87] focuses on reducing the overhead of QoS mechanisms. While using a priority based QoS approach, it tries to reduce the overhead by a topology aware QoS design. Kilo-NoC only provides QoS mechanisms in the parts of the network where needed and uses simple routers for the remainder. This approach is orthogonal to our approaches discussed in the remainder of this chapter. Hence, we can combine the ideas to only introduce our mechanisms to the routers that need to provide service guarantees, thus reducing the overall overhead of the QoS mechanisms.

Lee et al. [137] introduce *globally-synchronized frames* (GSF), for providing guaranteed QoS in terms of bandwidth and latency bounds. GSF coarsely divides the time into frames and introduces a scheme similar to *earliest deadline first* scheduling based on these frames. For this, each QoS packet from a source is assigned a frame number indicating the desired delivery time. Packets with an early delivery time are prioritized in the routers. Heißwolf et al. [93] propose to use a weighted round robin (WRR) scheduling policy. While GS traffic can use higher weights to obtain the needed QoS, it does not preempt best-effort traffic. The exclusively reserved communication resources allow to provide hard guarantees regarding throughput and latency.

Besides the prioritization of GS traffic, some approaches try to improve the performance of BE traffic. For this, BE can have the same or even a higher priority than GS traffic. To limit the interference, i.e., to still guarantee a minimum throughout for GS, mechanisms to dynamically adapt the priorities are introduced. Burns et al. [44] present *WPMC*, a protocol for priority-preemptive VC arbitration, guaranteeing that all (critical) packets will arrive by their deadlines. It uses runtime monitoring at the network interfaces (NI) to check whether critical traffic stays within a predefined behaviour (i.e. message sizes and inter-arrival times). If an injecting network interface (NI) detects a deviation from this behaviour, routers on the desired path switch to a critical state, in which all best-effort traffic is dropped by a router to favour the critical packets. This scheme is improved by Indrusiak et al. [103] to not drop best-effort traffic but allow it to use idling ports of a router even in the critical state. These schemes are similar to our approaches in Sections 4.3 and 4.4, as they allow prioritizing best-effort traffic over critical traffic, while monitoring is used to change the priority during run-time. The main difference is the monitored behaviour. In [44; 103] the behaviour of critical senders at the NI is monitored (i.e. message sizes and inter-arrival times), while our approaches monitor the interference packets of critical senders experience inside the network. Additionally, the monitoring at the injecting NI only allows to differ between critical packets, that are transmitted either with low or with high priority on the whole path [44; 103]. And after switching to the critical state, the routers remain in this state and henceforth prioritize critical traffic. Hence, the exploitation of the slack of critical applications, to increase the performance of best-effort traffic, is limited in the current design of [44; 103]. However, as these approaches monitor a different part of the NoC, they can be combined with ours to further exploit a system through monitoring the behaviour of critical tasks as well as the interference induced by non-critical ones.

In [59] the authors present distributed traffic shaping (DTS), where each output port uses a token bucket shaper. In this scheme, BE traffic is prioritized over critical traffic and the shapers are used to ensure that sufficient bandwidth remains for GS packets. However, this approach introduces a high overhead in the routers through the token bucket shapers. The authors of [58] introduce *backsuction*, in which BE is prioritized over GS traffic. In this scheme, rate limiters at the destinations nodes are used that return a control signal upstream along the transmission path. If a router recognizes a too low bandwidth of critical traffic, it increases the priority of the critical stream. This check is based on a threshold value in the routers, to denote when a buffer underflow occurs. This scheme enables the same performance benefits for BE as DTS [59] but can reduce the needed buffer size. As this approach uses a rate limiter at the destination node, it relies on the transmission of a leading (BE) packet to set up QoS in the path before the main transmission can start. Additionally, to correctly route the control signal upstream, this scheme allows only a single ongoing transmission in a (virtual) channel that uses the backsuction scheme (i.e. no sharing of virtual channels between multiple GS streams).

The authors of [224] introduce a *fluid meter* in each router, which is used together with a (3 bit) header extension denoting the bandwidth requirement of a packet/stream. Based on these two values a router can dynamically decrease or increase the priority of the packet. When GS requirements are satisfied, less VCs must be allocated to (multiple) GT streams (i.e. as they do not have to leave the router immediately). These VCs can be used by BE, to not sacrifice performance of BE. However, the approach increases the packet header and introduces an additional finite-state machine (FSM) in each router.

In summary, most of today's NoC architectures do not meet the requirements on isolation, flexibility, and high system utilization at the same time. TDM based architectures introduce static overhead due to the static time schedule, usually reducing the performance. Most of the dynamic QoS approaches favour safety-critical over best-effort (BE) traffic (e.g. strict prioritization), thus reducing the BE performance or introduce complex additional logic in the routers. However, the behaviour of the safety-critical applications is well-known and they do not benefit from finishing earlier than their deadline [201; 217]. Hence, they typically have some slack available, where the (latency) slack denotes the time budget between the worst-case latency of a packet and its deadline at the destination. This slack can safely be used to schedule other traffic [201; 217]. Hence, state-of-the-art QoS approaches typically overly degrade the performance of BE communication. To solve these problems, we propose methods that only introduce low overhead in the routers. They exploit the slack of safety-critical applications to increase the BE performance compared to other approaches, while still providing sufficient isolation.

4.3 Providing Efficient Latency Guarantees

In this section we describe a new approach providing latency guarantees while reducing the adverse effects of state-of-the-art QoS mechanisms on best-effort applications [219]. The goal is to exploit the latency slack of critical applications to increase the performance for best-effort traffic. Although the proposed mechanism is not specific to a certain network architecture, we restrict the explanations to a basic and commonly used architecture as shown in Section 2.1.6 and used, e.g., in [183; 216]). We assume a mesh network, where every router is connected to up to four neighbouring routers and one client (e.g. processing element or memory). The routers use wormhole switching and credit based flow control, i.e., buffer management is performed on equally sized flits, and have a four-stage pipeline. The input ports of the routers provide multiple separate virtual channels (VC) to prevent head-of-line blocking between certain traffic streams and thus to isolate different criticalities, i.e., there is no VC sharing between different criticalities. To reduce the size of the crossbar, each input can only send one flit from a single VC at a time over the crossbar and the routers use a two-stage arbitration. We use a "winner-takes-all" arbiter for requests of the same class, which is similar to a round-robin arbiter, but maintains a grant until the end of a packet. This improves average latency, as packets are sent in one piece if possible [54]. We do not consider speculation and look-ahead mechanisms like [130], which optimize the average latency. However, our approach can be used together with such techniques. In fact, as it reduces the contention latency of BE traffic, the relative effect of our approach can be even larger when combined with low-latency routers.

4.3.1 Baseline Architecture of the Approach

The baseline implementation of our GL approach divides the traffic into two different QoS classes: *guaranteed latency* (GL) and *best-effort* (BE). Each class uses a dedicated set of virtual channels and is assigned a distinct priority level. We use the GL channels for safety-critical traffic, while best-effort channels only contain non-critical traffic. If distinct priorities between different safety-critical applications are needed, more GL channels with different priorities can be used, similar as for standard priority based NoCs.

To enable the exploitation of the latency slack, we extend the packet header (for GL) with an additional field, which holds the slack information: the *blocking counter* (*BC*). This value is set in the NI and decremented in each router, based on the actual blocking experienced by the packet. The current BC value is then evaluated in each router to monitor the remaining slack time. This remaining slack time is used for a dynamic prioritization of GL traffic (cf. Section 4.3.2). In the baseline version, this counter can be on flit or packet level. Each time a GL packet is blocked by a higher priority stream, the counter is decremented by one. For a packet level counter this leads to some conservatism, as it is also decremented if GL is only blocked by a partial packet (e.g. the tail flit of a packet). If the network only supports a single packet size for all streams and this size is only a few flits, the packet level counter can be used to reduce the induced overhead, as fewer bits are needed in the header to store the counter. For networks with different packet sizes and long BE packets, it can be adverse for GL to be delayed by a full BE packet. Hence, the flit level counter enables a more fine granular capturing of the interference.

The use of the slack information in the header enables to freely distribute the allowed additional blocking over the network. This allows accounting for dynamics in the network, which typically lead to temporary, local traffic hot-spots. Additionally, it offers the ability to adapt the allowed blocking at the traffic source during runtime and to use different values for applications sharing a VC or even for packets of the same application. Thus, it also offers to integrate local monitoring information of the network interface [103; 217], or global information [118] in the allowed blocking to handle uncertainties of the system load. The initial value of the blocking counter is set in the network interface based on the results of a performance analysis (cf. Section 4.3.5).

4.3.2 Dynamic Prioritization

The arbitration logic in the routers selects the next request (i.e. VC) based on the priority of the VC. In our approach, the priority relation of the BE and GL class can be adapted dynamically by the routers based on the remaining slack time (i.e. current value of the blocking counter). We distinguish for each GL channel two different states in each router: *normal* and *critical*. In the *normal* state, the BE class has a higher priority than the GL channel. In the *critical* state, the GL channel obtains a higher priority than the BE class (and other GL channels in the normal state). With these priority levels and state definitions, the blocking counter of a GL packet in the *normal* state is decremented, each time the packet is blocked by a BE or a GL stream that is in the *critical* state. To determine the state of the VCs, a monitor checks the remaining slack time of the GL packets (i.e. state of the blocking counter). If there is any GL packet in a (input) queue, which has no more slack available, the state of the queue is set to *critical* until only packets with slack available remain. Even though this scheme has the drawback of prioritizing flits, that still allow more blocking (i.e. another packet at the queue head might still allow blocking, while the last packet requests for prioritization), it allows an easy implementation with a low hardware overhead. Additionally, for small buffer sizes the effect is reduced, as less flits can be in front of the packet that needs prioritization. Other possibilities are to move the packets that reached the maximum blocking to another queue (bypass) or to order the whole queue according to the remaining allowed blocking. This leads to a trade-off between increased BE performance and induced overhead when deciding whether a bypass, reordering mechanism, or simple queue prioritization should be used.

The arbitration of streams is packet based to increase the average performance of the system. Packet based arbitration also guarantees that the head and body flits of a packet experience the same additional blocking (e.g. the body flits cannot experience additional blocking through BE in a router after the header has been transmitted). Hence, the blocking counter must only be stored in the header flit. To prevent additional priority inversion after the slack is depleted, GL packets can preempt BE packets on flit level when being in the *critical* state.

4.3.3 Operational Example

This section provides a brief, simplified operational example for the arbitration to illustrate the core idea of the mechanism using a packet level blocking counter. In the *normal* state, the BE class has a higher priority than the GL channels. To limit the interference through the BE class on the GL channels, the routers dynamically change to the critical state based on the actual behaviour of the system. For this they monitor the blockage a GL packet experiences in the network. If the blockage reaches a certain threshold, the GL channel is prioritized over the BE class. Figure 4.2 shows a brief example for this approach using single flit packets.

The figure shows seven arbitration cycles of a simplified router in which two virtual channels (BE and GL) compete for the same output port. In cycle $t=0$, two packets are pending in the GL queue, while the BE queue is empty. Each GL packet is assigned the number of further allowed blocking, where both allow one additional blocking. Each time the packet is blocked by another packet with higher priority, this value is decremented. And when this value reaches zero, the GL channel obtains a higher priority. As both packets have a value greater zero, the state is *normal*. In the next cycle, the first GL packet is scheduled on the crossbar as no other packet was competing for it. At the same time, a new BE packet arrives. At cycle $t=2$, the BE and GL packets compete for the crossbar. As the state is normal, the BE packet has a higher priority and gets access to the crossbar. That way, it is not blocked by GL and can pass the router with a low latency. In the same cycle, a new BE packet arrives and the blocking counter of the GL packet is decremented, reaching zero and initiating a state change. In cycle $t=3$, the state is *critical*.

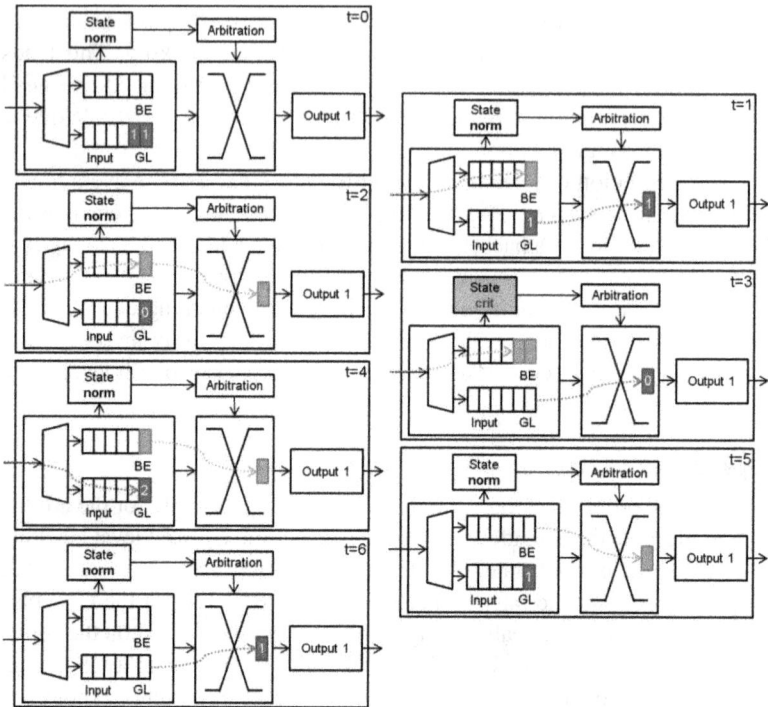

Figure 4.2: *Example for dynamic prioritization between a GL and BE chan-nel.*

Hence, the GL packet is forwarded and the BE queue is blocked. Again, a new BE packet arrives in this cycle. As there are no additional GL packets with a blocking counter value of zero, the state is changed to *normal* again. In the following cycle (*t=4*), one of the BE packets is forwarded, as no other packets compete for the crossbar. Additionally, a new GL packet arrives with a blocking counter value of two. In cycle *t=5*, the remaining BE packet is scheduled and the blockage of the pending GL packet is decremented. And finally, in the last cycle, the GL packet is forwarded, as no other packet competes for the crossbar.

As can be seen from the example, BE packets can pass the crossbar with a reduced latency, when the GL packets have a blocking counter value greater zero. GL traffic is only scheduled, when there are no other pending packets or the state is *critical*. Thus, as long as the GL packets are not blocked too often, BE can achieve a better performance.

4.3.4 Arbitration Logic

This section introduces an exemplary extension of the arbitration stage in the routers to implement our mechanism. In the arbitration stage, requests from all virtual channels and input ports must be processed along with their traffic class and a priority signal, denoting if the slack of a packet is used up. As we change the priority of the GL traffic according to the actual blocking in the network, we need an arbiter that can switch between different priorities for certain requests. Solutions for multi-priority arbiters are presented in [64; 232; 30, p. 11-27].

Figure 4.3 shows a simplified diagram of our arbitration stage of an output port. It consists of a request multiplexing and masking unit that assigns incoming *requests* from n input ports to the different class arbiters based on their *class* identifier. Additionally, it masks the requests for which backpressure is active and that do not have a *priority forward* (PF) signal. In the baseline implementation, all classes use a round-robin arbiter, while other arbitration policies are also possible. For the guaranteed latency class, we use two arbiters. One for the normal state (*GL norm*) and one for the critical state (*GL crit*). The two GL arbiters are used to individually keep track of the most recent senders on both priority levels, while other implementations, such as a single arbiter using two different states, are possible. If a GL request is raised, the request multiplexer sends it to the corresponding GL arbiters. A small qualifier logic then only allows requests that reached maximum blocking (*BC=0*) or have a *priority forward* signal to be assigned to the arbiter of critical GL requests. Additionally, this unit generates a *priority forward* (PF) signal when a GL sender reached maximum blocking but receives a backpressure signal. The priority forward signal is used to signal downstream routers that GL needs a higher priority. This covers the case, where a GL flit is blocked too often in a router but the preceding flits in the downstream router still allow blocking. In this case, the downstream router might raise a backpressure signal further delaying the GL flit. Next to the individual class arbiters, we have a static priority arbiter, sending out the highest priority grant to the requesting input ports.

Comparing the design with the arbitration logic of basic, priority-based routers, only small changes are needed. A simple router, using two different priorities for the virtual channels and round-robin arbitration between requests of the same priority, requires all blocks but the *qualifier*, the priority

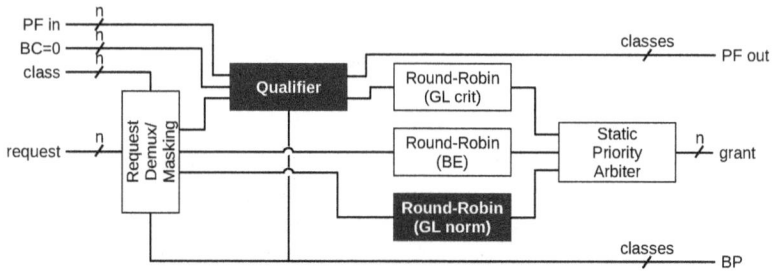

Figure 4.3: Simplified output port arbitration logic.

forward signal, and the additional (low priority) GL arbiter (state) [64; 232; 30, p. 11-27]. Hence, our approach only extends an additional round-robin arbiter (or state), a priority forward signal, and the qualifier logic, which is mainly composed of *and*-gates forwarding only requests for which the slack is used up.

4.3.5 Analysis of the Approach

In this section we present an analysis for the upper latency of GL traffic using the compositional performance analysis (CPA) framework (cf. Section 3.4) for our mechanism, proving sufficient independence. To improve readability, we restrict the analysis to the case where all GL streams share a single VC. However, it can straightforwardly be extended to handle multiple virtual channels and additional priority levels (cf. Section 3.4). For the analysis according to Section 3.4, we need to derive the corresponding worst-case multiple activation processing time $B_i^+(q, a_i^q)$ of a stream in a router. It denotes the maximum time required to transfer q flits of a stream i, given that all but the first flit arrive before their respective predecessor has been transferred and the q-th flit arrives at time instant a_i^q. Based on this, we then derive metrics (e.g. path latency) for a single router and for a complete network.

To conservatively capture all possible worst-case scenarios, we need to extend the multiple activation processing time from Section 3.4 by the additional interference of the mechanism. We denote this as **dynamic priority blocking** $B_{i,q}^{dyn}$, covering the (remaining) amount of time stream i is blocked being in the normal state.

With this we can adapt the inequality from Equation 3.3. As all GL senders share a single channel, we can remove some bocking factors. Ad-

ditionally, the new arbitration scheme for GL introduces a new one (B^{dyn}).
Hence, the upper limit for the multiple activation processing time can be
derived as:

$$
B_i^+(q,a_i^q) \leq q \cdot C + B_i^{out,ep}(B_i^+(q,a_i^q) - C + \varepsilon, q) + B_i^{fifo}(B_i^+(q,a_i^q), a_i^q)
$$
$$
+ B_{P(i),VC(i)}^{bp}(q) + B_{i,q}^{dyn}(B_i^+(q,a_i^q) - C) + B^{lp} . \tag{4.1}
$$

For the worst-case, we assume full blocking while being in the normal state
(B^{dyn}), followed by the maximum blocking from other GL streams on the
router ($B^{out} + B^{fifo}$) and backpressure (B^{bp}). As all GL senders share a single
channel and B^{dyn} accounts for all blocking while being in the normal state,
there can be no blocking through any higher priority streams. Hence, there
is no input, indirect output blocking, or higher priority output blocking
(compared to Theorem 3.4.1). To derive the multiple activation processing
time on a single router, we now derive the individual sources of blocking.

Theorem 4.3.1 The output blocking $B_i^{out,ep}$ in the considered example that
a GL stream i may experience in any time window consists of the blocking
through other GL streams from other input ports. It can be bounded by:

$$
B_i^{out,ep}(\Delta t, q) \leq \sum_{p \in Out_i^P} \left(C \cdot \hat{\chi}_p + B_{P(i),VC(i)}^{bp}(\hat{\chi}_p) \right)
$$
$$
with \quad \hat{\chi}_p = \min \left\{ \left(\left\lceil \frac{q}{n} \right\rceil + 1 \right) \cdot n, n \cdot \hat{\rho}(\Delta t, p, OUT(i), VC(i)) \right\},
$$
$$\tag{4.2}$$

where n is the packet size, $VC(i)$ and $OUT(i)$ denote the virtual channel
and output of stream i, and Out_i^P is the set of other input ports.

Proof. The proof is analogous to the proof for Lemma 3.4.3. As we already
accounted for the full blocking being in the normal state in Equation 4.1
($B_{i,q}^{dyn}$), we only need to account for GL streams from other input ports. And
with the definition of the arbitration and output blocking, there can only be
blocking of other GL streams on the same VC.

Due to wormhole switching, once the scheduler grants access to an output
port VC, no other input port can access this output port VC until it is released,
i.e., the packet is fully transmitted. This is captured by $\hat{\rho}$, which considers
that after a head flit arrives within the time interval Δt, the whole packet will
be served before stream i. Additionally, due to the round-robin arbitration,

each head flit belonging to stream i may only be blocked once by each other input port. But as the state of the GL sender, and hence also the arbiter state, might have changed, an additional round-robin cycle can occur. This is addressed with the *min*-function, where $\left\lceil \frac{q}{n} \right\rceil$ is an upper bound on the number of head flits and the plus one denotes the additional round-robin cycle. Each of these head flits can be blocked at most for n flits from each other input port. Moreover, each of the interfering flits then will block stream i for the flit transfer time C and the backpressure blocking these flits experience. ∎

Theorem 4.3.2 The FIFO blocking B_i^{fifo} in the considered example of a GL stream i in any time window consists of the blocking of other GL streams that share the same input port (and VC). It can be bounded by:

$$B_i^{fifo}(\Delta t, a_i^q) = m \cdot C + \max_{\theta \in \Theta^k} \{A_\theta\} + \max_{\theta \in \Theta^l} \left\{ B_{P(\theta),VC(\theta)}^{bp}(m - k \cdot n) \right\}$$

$$\text{with} \quad m = \min \left\{ \left\lceil \frac{a_i^q}{n} \right\rceil \cdot n, \sum_{j \in Buf_i} \left\{ \rho_j^+(a_i^q) \right\} \right\}$$

$$k = \left\lfloor \frac{m}{n} \right\rfloor$$

$$A_\theta = \sum_{j \in \theta} \left\{ B_j^{out,ep}(\Delta t - C, n) + B_{P(j),VC(j)}^{bp}(n) \right\}, \tag{4.3}$$

where Buf_i denotes the set of all streams sharing the buffer of stream i; and k denotes the maximum number of whole packets (and hence head flits) of other streams.

Proof. The proof is analogous to the proof for Lemma 3.4.4. The blocking caused by other streams at the same input consists of the transmission time of the flits that arrived before the q-th flit of stream i and the interference those flits observe. The first term accounts for the transmission of these flits. For this, only flits that arrived before the arrival of the q-th flit can be in the buffer. Thus, m provides the maximum number of flits of other packets that can be in front the q-th flit of stream i in the buffer.

These flits may also observe blocking that influences the q-th flit of stream i. Here, the interfering flits can be grouped into flits of whole packets as well as a packet partially transmitted at the front of the queue. The second term accounts for the worst-case blocking whole packets can observe. It considers for the k packets all possible mappings to streams and takes the maximum blocking. This blocking consists of the output and backpressure blocking each packet will experience. And the third term accounts for the blocking of a partial packet. As the header of this packet has already been sent, we only need to account for backpressure blocking.

As streams i and j share the same interferer set at the input port, all FIFO blocking of stream j is accounted for in the FIFO blocking of stream i and must not be accounted here again. ∎

Theorem 4.3.3 The blocking while being in the normal (i.e. low priority) state $B_{i,q}^{dyn}$ in the considered example of the q-th flit of a stream i can be derived from the blocking counter (BC) in the flit header. It can be denoted as:

$$B_{i,q}^{dyn}(\Delta t) \leq C \cdot \min \left\{ \sum_{j \in hp(i)} \eta_j^+(\Delta t), \kappa_i(\Delta t) \right\}$$

$$with \quad \kappa_i = \begin{cases} BC_i^q \cdot \hat{n}, & \text{if } BC \text{ counts packets} \\ BC_i^q, & \text{otherwise} \end{cases} \tag{4.4}$$

where BC_i^q denotes the current blocking counter value of the q-th flit and $hp(i)$ the set of all interfering streams for stream i that might have a higher priority (i.e. BE and critical GL) at the router under analysis.

Proof. From Section 4.3.1 we know that the flit only is in the normal state as long as the blocking counter has slack available. Depending on the implemented granularity (packet or flit based counter) it is decremented for each packet or flit with a higher priority (BE or critical GL) that blocks stream i. When the counter reaches zero, no more blocking is allowed. Hence, only BC_i^q packets or flits can interfere with the q-th flit of i. Additionally, we only have to account for interfering flits that can arrive during the time the q-th flit is waiting at the router, covered by the min-function and the sum of arrivals of all interfering streams (η_j^+). ∎

With all sources of blocking defined, the multiple activation processing time from Equation 4.1 for a single router is fully defined. Based on this we can derive an upper limit for the worst-case single hop latency and end-to-end metrics for the whole network as discussed in Sections 3.4.3 and 3.4.3.

In addition to these metrics, we need to extend the path latency and derive the maximum dynamic priority blocking that a GL packet allows in each router on its path (i.e. the current value of BC). A naive approach is to assume the initial value in all routers, which leads to pessimistic but valid results. The pessimism can be reduced, if we can assume a certain minimum and maximum load from the interfering streams, such that there is always some or no dynamic priority blocking for a stream i in the normal state and derive more accurate propagation models for possible values. If there is a minimum load through BE traffic on a router, some dynamic priority blocking will always occur. And if there is an upper bound on the interfering load, not more than this can occur. This information can be used to reduce the set of possible combinations. For example, if we can assume the full initial blocking value for the first router on a path, we know, that there will be no additional dynamic priority blocking on the subsequent routers as the stream switched to the critical state. Hence, to optimize the analysis results, we then can analyse all possible distributions of the blocking counter on the path p of a stream i and derive the maximum end-to-end latency $l_p^+(q)$ for transmitting q flits from all distributions. However, this exhaustive search checks $\binom{hops+BC-1}{BC}$ combinations. Following this, the maximum end-to-end latency consists of the worst-case response time for each hop on the path p that depends on the assumed value for BC, the time to inject the q flits, and the packetization overhead:

$$l_p^+(q) \leq \max\left\{\delta_{First(p)}^-(q), \hat{B}_{P(First(p))}^+ \left(q+b_{P(First(p))} - Q_b\right)\right\}$$

$$+O_p + \max_{d \in Dist}\left\{\sum_{j \in Tasks(p)} R_{j,d}^+\right\}, \tag{4.5}$$

where $First(p)$ defines the first task of the chain (i.e. network path); $Tasks(p)$ the set of all tasks of path p (i.e. one per hop), O_p the constant de/packetization overhead; $R_{j,d}^+$ the worst-case single hop latency assuming a distribution d of the dynamic priority blocking; $\delta_{First(p)}^-(q)$ denotes the time the sender needs to inject q flits (assuming no contention); Q_b is the size of the buffer; $\hat{B}_1^+ (q+b_i - Q_b)$ the overhead induced by backpressure until the q-th flit and all backlogged flits of previous transmissions can be injected to the first router; and $Dist$ the set of all possible distributions of the dynamic priority blocking for the stream p. Basically the equation computes how long it takes a stream to inject q flits and then assumes the last one of these to experience the worst-case blocking on all intermediate routers. For this it checks all possible distributions of the allowed additional blocking through BE streams. Due to the in-order delivery of the network, all previous flits will have arrived at the destination before the last one. And the delay previous flits may observe is included as interference in the worst-case blocking of the last flit.

4.3.6 Finding Admissible BC Values

The analysis above provides all means to find admissible values for the allowed dynamic priority blocking (i.e. for the initial blocking counter value) for a GL stream to exploit its slack. For this, we can define an iterative approach that compares the worst-case end-to-end latency against the deadline of a stream to derive its slack. First, we analyse the system, assuming all blocking counters to be zero (i.e. classic prioritization of GL) and the system to be schedulable. Then we identify the GL streams that have slack by comparing their derived worst-case latency with their deadline. If any stream has slack, we can increase the initial BC value of this stream and re-analyse the system to check whether all deadlines are still met. In this step, all streams (or more precisely all streams that are directly or indirectly interfered by the stream under consideration) need be to checked. This is needed as the modified BC value can lead to changed output event models for all streams on routers that are used by the stream under consideration. Additional, these changed event models can propagate to other routers, and thus influence further streams. Doing this, we can find feasible initial values of dynamic priority blocking for all streams. Besides this simple approach, more complex strategies for defining initial BC values are possible. For example, the approach presented by Indrusiak et al. [103] can be used to define BC values for different sender states based on the core- or NI-local behaviour (e.g. message sizes and inter-arrival times). That way, global and node-local information cab be used to derive a feasible value for BC and adapt it during run-time.

4.3.7 Evaluation

In this section, we evaluate the performance of the GL mechanism and compare it against the classic and widely used static prioritization [39; 87]. We divide the evaluation into two parts. In the first part, we use synthetic workloads to evaluate the basic functioning and certain properties of our mechanism, such as isolation between BE and GL and the correctness of the analysis. In the second part, we use memory access and communication traces of general purpose applications, to investigate the performance of the mechanism on realistic workloads. All experiments were carried out with the *OMNeT++* simulation framework and an extension of the *HNOCS*

library [32] (cf. Section 2.3) using routers with a four-stage pipeline, buffers
to store six packets in each virtual channel of each input port, and a packet
size of four flits.

Synthetic Workloads

The first set of experiments uses synthetic workloads, generated based on
average link loads, and the simple system shown in Figure 4.4. It comprises
a line topology to maximize the overlap between traffic streams. In the
example, we have five traffic streams, each periodically injecting packets
with a jitter of 25 % of its packet injection period. There is a guaranteed
latency (GL) communication from task τ_1 on the first node to τ_7 on the last
node. In between there are four best-effort (BE) communications, where
each BE task sends data to its direct neighbour on the right. This example
is compact enough to be comprehensively displayed but shows all relevant
effects of the mechanism.

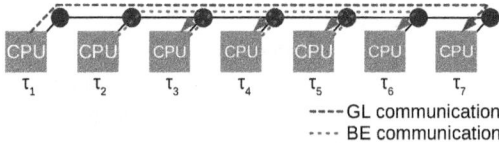

Figure 4.4: *Simple communication scenario using a line topology.*

In a first experiment, we set the GL task to generate an average network
load of 10 % and increase the load of the BE tasks (i.e. reduce the period
between packets), to measure their interference on GL. For GL, we evaluate
four different values for the allowed dynamic priority blocking (BC) on flit
level granularity. Figure 4.5 shows the end-to-end latency of full packets for
the GL sender in this experiment. In the figure solid lines show the measured
maximum latency and dashed lines the results of the worst-case analysis
for GL assuming maximum BE load. If we allow zero dynamic priority
blocking (BC=0), GL traffic always has a higher priority than BE traffic,
which corresponds to classic prioritization of GL. In this case, an increasing
BE load does not influence GL traffic. That is, the GL latency is low and
constant over an increasing BE load.

With the allowance of LP blocking ($BC > 0$), BE traffic can interfere with
GL traffic, where a higher BC value leads to a higher possible interference.
However, the interference is upper bounded and thus from a certain BE load
onwards, the GL latency is not further increasing. In all cases, the observed

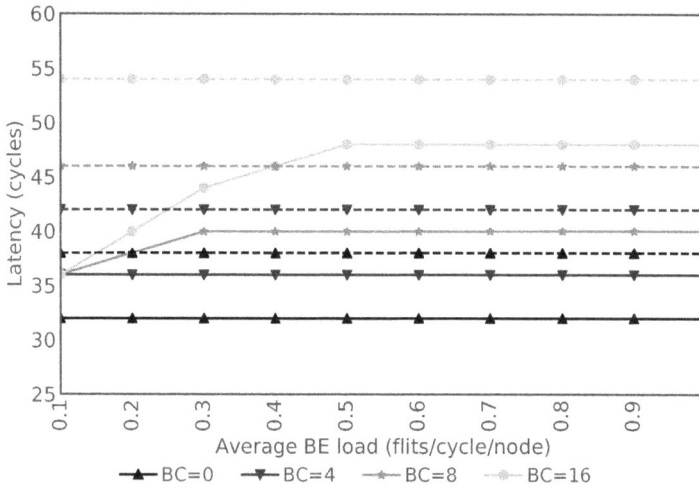

Figure 4.5: *Measured (*solid*) and analyzed (*dashed*) worst-case packet latency of GL for synthetic BE loads and various blockage values.*

upper bound is below the analysis results. The results also show that for small loads the observed latency for GL is similar for different BC values. This is due to the fact that the possibility for GL and BE flits to compete in a router for resources depends on the load in the system. With a low load, the flits rarely compete in the routers, and hence the worst-case might not be observed in simulation. For high loads, however, the flits compete more often, leading to a higher interference and thus latency, especially for higher values of BC.

In a second experiment, we investigate the influence of GL on BE traffic. For this, we compare the classic prioritization of GL traffic against our approach for a BE load of 20 %. Figure 4.6 shows the average latency results of this experiment for the BE tasks τ_2 and τ_5. For classic prioritization of GL (denoted as *HP*), the latency for both best-effort tasks behaves the same. With an increasing GL load, the latency increases, due to the higher priority of GL. The latency of BE tasks thus depends significantly on the load of GL traffic. For the new approach, the experiment shows a better performance for BE tasks and loosens the dependency on GL load.

Additionally, the experiment shows that the value of the allowed dynamic priority blocking and the distance to the GL sender influence the performance

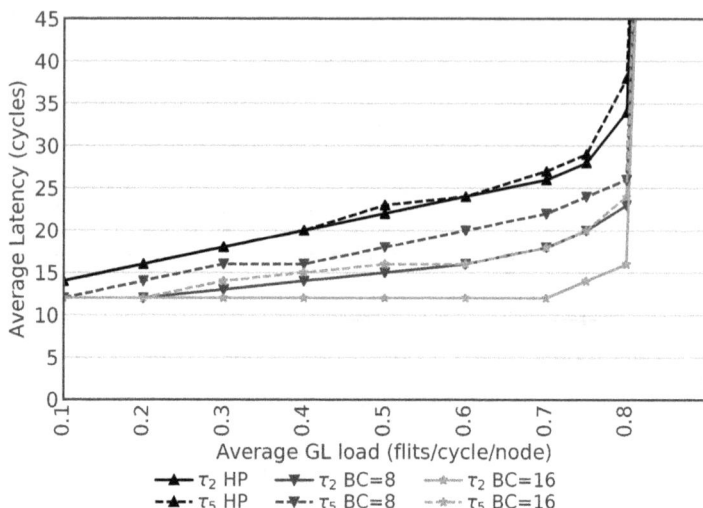

Figure 4.6: *Average BE latency for synthetic GL loads.*

benefit of BE tasks. Here the task τ_2 has a better performance than τ_5. This results from the fact that for τ_2 the allowed dynamic priority blocking of the GL flits is still at its initial value. Thus, flits from τ_2 can pass the router with a higher priority than GL, even for increased loads. With an increasing distance, the remaining allowed dynamic priority blocking tends to be lower. Thus, the task τ_5 has a higher probability to be blocked by GL flits and to experience a higher latency.

With the allowance of blocking through BE, the GL packets also stay longer in the router, which can increase the backlog. Figure 4.7 shows the backlog derived from the analysis for the different configurations for the same experiment. As can be seen, the backlog depends on the GL load and BC value. Especially for high loads and high values of BC the backlog can increase drastically. To prevent this, rate limiters at the source can be used, which are available in many existing NoCs, or the analysed backlog is used to make the buffers big enough.

Benchmark Workloads

In the second set of experiments we use realistic workloads to evaluate the performance of the proposed mechanism. For the experiments we obtained traces from the CHStone benchmark suite [91]. The traces were extracted

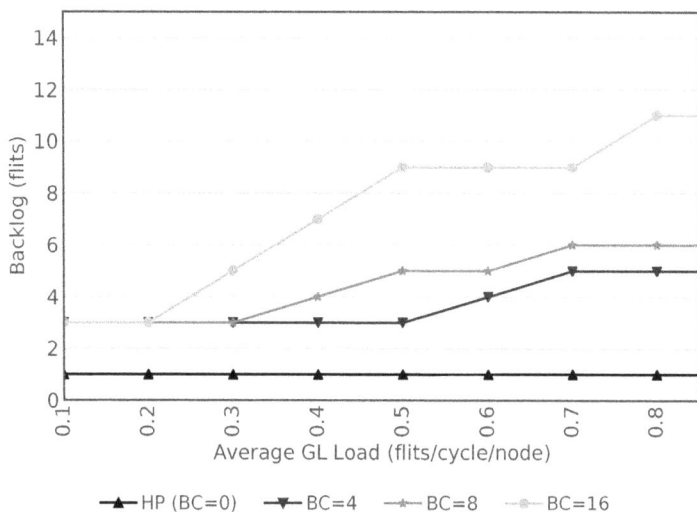

Figure 4.7: *Worst-case backlog of GL derived from the analysis.*

using the Gem5 simulator and an ARMv7-a core with a 32 kB L1 cache. Each trace contained 100000 accesses to the network, where each access can be a direct memory access, communication, or a cache access. The compilation was performed using a standard *gcc* compiler (ver. 4.7.3). For a simulation run, the different traces were randomly assigned to nodes in a *4×4* mesh network as shown in Figure 4.8 to obtain different traffic distributions, while the nodes and the network were running with the same clock frequency. In the network, we assign fixed QoS classes to certain nodes. There are five nodes initiating GL traffic and eleven nodes for BE traffic. Additionally, there is one memory with a single interface to the network. For the routing policy we use an XY-routing. Thus, when regarding the highlighted BE node, all GL flits accessing the memory will compete with flits of this node. This scenario is compact enough to be comprehensively displayed but shows all relevant effects of the mechanism. Based on this scenario, we conduct two series of experiments. In the first series we use the memory as a hot-module. That is, all traffic generated by the nodes is sent to the memory, leading to higher interference on links near the memory. In the second series we use pseudo random destinations for the traffic of all nodes.

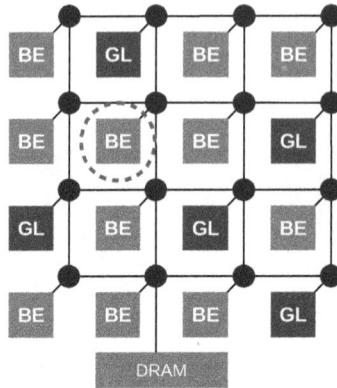

Figure 4.8: Communication scenario for benchmark applications.

Figure 4.9 presents the results of the first experiment, with the memory as the hot-module. It shows a box plot of the measured latencies for eight different applications mapped to the highlighted node. For each run, the box covers 50 % of the latencies, with its lower and upper borders giving the 25 % and 75 % quartiles. The whiskers indicate the measured minimum and maximum observed latency. The median and average among the measured latencies are respectively marked by a black bar and a black dot. For each of these applications, we generated 1000 different sets of interfering workloads for the other nodes (i.e. random mappings of the benchmarks to the network nodes). The results are presented for three different prioritization approaches of the GL nodes: the classic prioritization *(HP)* and the new approach with two different values for the allowed dynamic priority blocking on flit level *(BC=4)* and *BC=8).*

The results show that the new approach reduces the average latency for both configurations compared to classic prioritization by up to 36 % for a blockage value of four and up to 43 % for eight. The minimum and maximum latencies are similar for all schemes. For a single hot-module, the different GL flits experience high interference on the way to the hot-module. Hence, it is likely that the maximum allowed blocking can be consumed, leading to a worst-case blocking for BE. On the other hand, the dynamic behaviour of the applications also leads to cases, where flits pass the network with nearly no interference, leading to the best-case latency. In this experiment, our approach increased the backlog of GL from 16 flits using classic prioritization to up

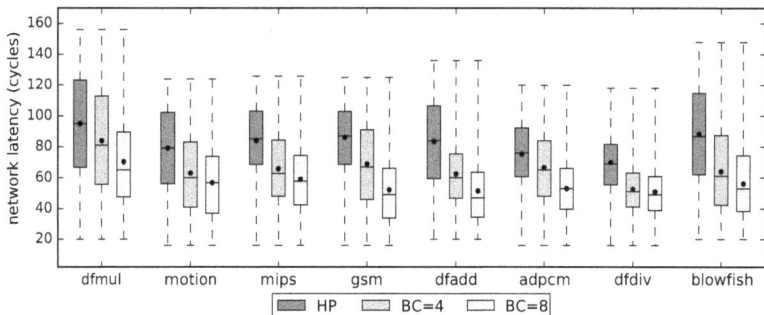

Figure 4.9: *Performance of CHStone benchmarks as BE traffic with memory as hot-module.*

to 20 flits with $BC = 8$. Compared to the synthetic workloads, the backlog of GL is higher even for classic prioritization, as multiple GL senders now share a virtual channel.

Figure 4.10 shows the normalized latencies as box plots for the system using pseudo random destinations for the traffic. We generated for each listed benchmark 1000 different sets of interfering workloads (i.e. random assignment of applications to nodes) and a random destination for each sender. We selected the destinations such that a traffic stream has to pass at least three routers (i.e. no traffic to direct neighbours). During a single run, an application was always sending data to the same node. This simulates scenarios, where applications exchange data or multiple memories exist. For the figure, we normalized for each run the latency to the average latency when using classic prioritization for GL (*HP*).

Again, the results show that our approach leads to a performance increase for the BE applications compared to classic prioritization of GL traffic. If we allow four blockages through the best-effort class, we can reduce the average BE latency by up to 30 %. And when we increase the allowed blockage to eight, we can reach a latency reduction by up to 45 %. However, the achievable reduction of the latency depends on the application behaviour. At the same time, the approach also increases the variability in the occurring latencies and backlog. With an allowed blockage of $BC = 8$ the backlog increases from 11 to 14 flits compared to classic prioritization. This increase is smaller than for the case of the hot module, as with random traffic less GL senders compete for the same buffer.

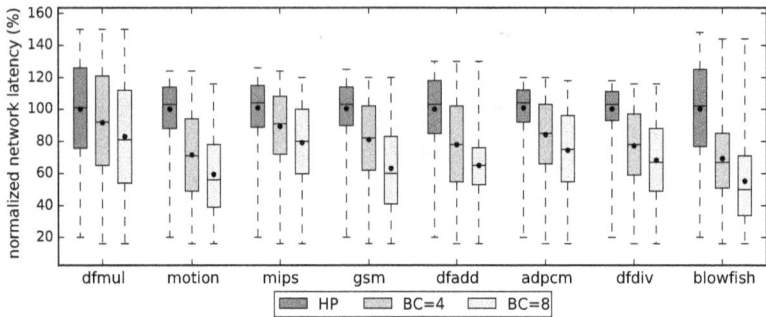

Figure 4.10: *Performance of CHStone benchmarks as BE traffic for random destinations normalized to average latency at classic prioritization.*

Synthesis Results

In this section we briefly present synthesis results for our approach. For this we implemented and synthesized a *2×2* NoC on a *Virtex-6 LX760 FPGA* using *Xilinx ISE 14.6* with default optimization settings and no special optimizations for the *VHDL* implementation. The device utilization data were collected from the *Module Level Utilization Summary Report* produced by ISE. As this NoC is not fully connected, each of the four routers has only three input ports fully instantiated. The results for the whole NoC are summarized in Table 4.1. The table compares the used *registers*, *LUTs*, and achievable clock frequency for four different implementations. The *baseline* implementation corresponds to a basic round-robin router with five virtual channels (VCs) and a buffer depth of six packets for each VC. This was extended in *FP* to provide one prioritized VC, e.g., *VC0*. In the *DP* design the priority of *VC0* can be changed via a configuration flag from the highest to the lowest priority during run-time. And finally, the *BC* implementation denotes our approach, where the priority of *VC0* is dynamically changed by the router based on the current blocking counter value in the flit header.

We used 4 bit to store the blocking counter in the header that were previously unused. If no spare bits are available in the header, the header, and possibly the signal width between the routers, must also be extended, resulting in additional overhead. The synthesis shows that the new approach introduces less than 5 % overhead for the used *2×2* NoC. This corresponds to an approximate increase of 1.67 % of a router per instantiated port. However, if no source rate limiters are used, our approach can increase the worst-case

Table 4.1: Synthesis results of a 2×2 NoC on Virtex-6 LX760 FPGA.

Unit	Baseline	FP	DP	BC
#Registers	5749	5767	5764	5980
#LUTs	7391	7425	7422	7719
Frequency (MHz)	210	210	210	210

backlog (e.g. from 11 to 14 flits in the benchmark example) and hence might require bigger buffers (if no backpressure aware analysis is used or the propagation of blocking leads adverse interdependencies) and further increasing the overhead. The achievable clock frequency was the same for all designs, showing that the extensions did not influence the critical timing path. The frequency was restricted by the minimum achievable period, caused by a *data path delay* of 4.75 ns, consisting of 1.31 ns for logic and 3.44 ns route delay.

4.4 Providing Efficient Throughput Guarantees

This section describes a new approach providing throughput guarantees while reducing adverse effects of state-of-the-art QoS mechanisms on non safety-critical applications. The goal is to exploit the throughput and latency slack of safety-critical guaranteed-throughput (GT) traffic to increase the performance for best-effort (BE) traffic. For this, we give priority to BE traffic for optimal latency and at the same time monitor the progress of GT traffic, to change priorities if needed. Although the proposed mechanism is not specific to a certain network architecture, we restrict the explanations to the same basic and commonly used architecture as discussed in Section 4.3.

The key idea of our approach is to prioritize BE traffic for optimal latency and at the same time monitor the progress of GT traffic, to increase the priority of GT if needed. For this, the proposed mechanism comprises up to four elements:

1. a *progress monitor* supervising GT,
2. a *selective priority arbiter* that uses the progress monitor for GT (i.e. the current flit buffer levels) as a decision criterion in the input stages of each network switch,
3. an entity at the source that tags the last packet of a GT connection, and
4. a rate limiter at the source.

If a GT sender is guaranteed to behave as specified (e.g. through the design process), the rate limiters for GT are not needed.

4.4.1 Progress Monitor and Selective Priority Arbiter

The arbitration logic is located at each output port (cf. Section 2.1). It processes the requests from all input ports according to their class signal and the current state of GT. Figure 4.11 shows a simplified block diagram of the arbitration logic for our approach. In the figure, t_n denotes whether an input port with GT traffic needs a higher priority. Based on this signal, a request of a GT stream is forwarded to the corresponding GT arbiter (i.e. *critical* when above the threshold or *norm* when beneath). And the *class* signal decides, whether the request results from a BE or GT stream. The *static priority arbiter* then selects the highest priority signal that has requests pending. If t_n is not asserted, the arbiter prioritizes BE, while GT can use empty slots (i.e. not used by BE).

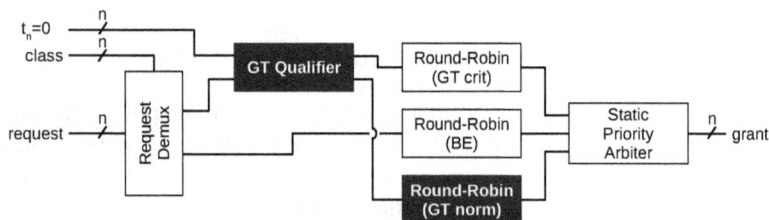

Figure 4.11: *Selective priority arbiter for guaranteed throughput traffic.*

The priority signal t_n is derived by the *progress monitor*. The progress monitor checks the buffer fill level or the presence of a special tail flit, named *end of transmission (EoT)* flit. For this, the routers have a (configurable) threshold value. When the buffer fill level is above the threshold or the EoT flit is in the buffer, the priority signal is asserted. Routers typically already have a measure for the fill level for the buffers for flow control. We only extend it by an additional threshold, which is then used to construct a priority feedback signal. This threshold can be configurable or static.

In summary, the proposed arbiter selects BE traffic, as long as all (requesting) GT channels are beneath their threshold and there is no EoT flit pending, and a GT request otherwise. If there are no requests of a specific priority, requests from a lower priority are selected. This means that GT is allowed to send if the link is otherwise idle, but also enables BE traffic to use unused reserved GT throughput, avoiding waste of over-allocated throughput.

Contrary to [58], this scheme allows multiple GT streams to share the same VC, as the priority signal must not be forwarded upstream, but is derived locally in each router. However, this also allows for possible head of line blocking between GT streams. Hence, only streams where the sink is known to accept traffic should be allowed to share a VC, which can be guaranteed by design or the use of a control layer (cf. Section 4.5) [125; 223].

4.4.2 Sender Extensions

For the proper work of the arbitration in the routers, the senders must be equipped with the possibility to tag the end of a GT transmission with the EoT flit. Additionally, if the behaviour of GT senders cannot be guaranteed by design, rate limiters are needed. The EoT flit is needed to enable the raise of the priority even when the buffer has not reached the threshold. This is needed, as when a GT streams sends its last packet and there are no other GT stream sending data on the shared path, there will be no flits accumulating in the buffer. Booth mechanisms can be implemented in software or as hardware extensions in the network interface (NI). As backpressure might occur at the injecting interface and router, the source needs sufficiently sized buffers or a stateful rate-limiter. This is needed to catch up a possible backlog at the sender with a temporary higher rate than requested. A simple example for a stateful rate-limiter, in this sense, is a token bucket shaper, where the bucket size covers the worst case backlog [138; 147; 209]. The bucket size then allows a burst, where the sender obtains more throughput than initially requested, to catch up an initial too low accepted throughput. This enables a *Super-GT* [146] like service, where a GT stream can obtain more traffic than initially requested in the long term, to recover from blocking.

4.4.3 Analysis of the Approach

In this section we show that our GT approach can guarantee a minimum accepted traffic rate for a sender. For this, we derive a lower bound on the minimum service and an upper bound on the backlog at the sender. In general, these values can be obtained using any analysis framework, such as [111; 134; 177; 220]. In the following, we will utilize the approach from Section 3.4, as it can handle backpressure for arbitrary sized buffers, to obtain the minimum accepted throughput for a sender and its backlog. To improve readability, we restrict the analysis to the case where all GT streams share a single VC and only present the equations that will change.

To derive the worst-case accepted traffic of a sender, we need to obtain the worst-case waiting time $\hat{B}_p^+$ of a sender at each router. As the waiting time depends on the traffic models of other streams, we also need to derive the response times and thus the multiple activation processing time of each stream. To conservatively capture all possible worst-case scenarios, we need to extend the multiple activation processing time from Section 3.4 by the additional interference introduced by the mechanism similar to the case for the GL approach in Section 4.3.5. We denote this as **dynamic priority blocking** $B_{i,q}^{dyn}$, covering the amount of time stream i is blocked being in the normal state.

With this we can adapt the inequality from Equation 3.3. As all GT senders share a single channel, we can remove some bocking factors. Additionally, the new arbitration scheme for GT introduces a new one (B^{dyn}). Hence, the upper bound for the multiple activation processing time can be derived as:

$$
\begin{aligned}
B_i^+(q,a_i^q) \leq &\, q \cdot C + B_i^{out,ep}(B_i^+(q,a_i^q) - C + \varepsilon, q) \\
&+ B_i^{fifo}(B_i^+(q,a_i^q),a_i^q) \\
&+ B_{P(i),VC(i)}^{bp}(q) + B_{i,q}^{dyn} + B^{lp} .
\end{aligned} \tag{4.6}
$$

For the worst-case, we assume full blocking while being in the normal state (B^{dyn}), followed by the maximum blocking from other GT streams on the router ($B^{out} + B^{fifo}$) and backpressure (B^{bp}) as defined in Definition 3.4.5. As all GT senders share a single channel and B^{dyn} accounts for all blocking while being in the normal state, there can be no blocking through any higher priority streams. Hence, there is no input, indirect output blocking, or higher priority output blocking. To derive the multiple activation processing time on a single router, we now derive the individual sources of blocking.

Theorem 4.4.1 The output blocking $B_i^{out,ep}$ that a GT stream i experiences in the considered example in any time window consists of the blocking through other GT streams from other input ports. It can be bounded by:

$$B_i^{out,ep}(\Delta t, q) \leq \sum_{p \in Out_i^P} \left(C \cdot \hat{\chi}_p + B_{P(i),VC(i)}^{bp}(\hat{\chi}_p) \right)$$

$$with \quad \hat{\chi}_p = \min\left\{ \left(\left\lceil \frac{q}{n} \right\rceil + 1 \right) \cdot n, n \cdot \hat{\rho} \left(\Delta t, p, OUT(i), VC(i) \right) \right\},$$

(4.7)

where n is the packet size, $VC(i)$ and $OUT(i)$ denote the virtual channel and output of stream i, and Out_i^P is the set of other input ports.

Proof. The proof is analogous to the proof for Lemma 3.4.3. With the definition of the arbitration and output blocking and as the additional delay while being in the low priority state is already covered in Equation 4.6, there can only be blocking of other GT streams on the same VC.

Due to wormhole switching, once the scheduler grants access to an output port VC, no other input port can access this output port VC until it is released, i.e., the packet is fully transmitted. This is captured by $\hat{\rho}$, which considers that after a head flit arrives within the time interval Δt, the whole packet will be served before stream i. Additionally, due to the round-robin arbitration, each head flit belonging to stream i may only be blocked once by each other input port. But as the state of the GT sender, and hence also the arbiter state, might have changed, an additional round-robin cycle can occur. This is addressed with the *min*-function, where $\left\lceil \frac{q}{n} \right\rceil$ is an upper bound on the number of head flits and the plus one denotes the additional round-robin cycle. Each of these head flits can be blocked at most for n flits from each other input port. Moreover, each of the interfering flits then will block stream i for the flit transfer time C and the backpressure blocking these flits experience. ∎

Theorem 4.4.2 The FIFO blocking B_i^{fifo} of a GT stream i in the considered example in any time window consists of the blocking of other GT streams that share the same input port (and VC). It can be bounded by:

$$B_i^{fifo}(\Delta t, a_i^q) = m \cdot C + \max_{\theta \in \Theta^k} \{A_\theta\} + \max_{\theta \in \Theta^l} \left\{ B_{P(\theta),VC(\theta)}^{bp}(m - k \cdot n) \right\}$$

$$with \quad m = \min \left\{ \left\lceil \frac{a_i^q}{n} \right\rceil \cdot n, \sum_{j \in Buf_i} \left\{ \rho_j^+(a_i^q) \right\} \right\}$$

$$k = \left\lfloor \frac{m}{n} \right\rfloor$$

$$A_\theta = \sum_{j \in \theta} \left\{ B_j^{out,ep}(\Delta t - C, n) + B_{P(j,VC(j))}^{bp}(n) \right\}, \tag{4.8}$$

where Buf_i denotes the set of all streams sharing the buffer of stream i; and k denotes the maximum number of whole packets of other streams.

Proof. The proof is analogous to the proof for Lemma 3.4.4. The blocking caused by other streams at the same input queue consists of the transmission time of the flits that arrived before the q-th flit of stream i and the interference those flits observe. As the GT stream waits until the threshold value is reached (covered by $B_{i,q}^{dyn}$ in Equation 4.6), all flits in the queue have the highest priority. The first term accounts for the transmission of these flits. For this, only flits that arrived before the arrival of the q-th flit can be in the buffer. Thus, m provides the maximum number of flits of other packets that can be before the q-th flit of stream i in the buffer.

Then, these flits may also observe blocking that influences the q-th flit of stream i. The interfering flits can be grouped into flits of whole packets as well as a packet partially transmitted at the front of the queue. The second term accounts for the worst-case blocking whole packets can observe. It considers for the k packets all possible mappings to output ports and takes the maximum blocking. This blocking consists of the output and backpressure blocking each packet will experience. And the third term accounts for the blocking of a partial packet. As the header of this packet has already been sent, we only need to account for backpressure blocking.

As stream i and j share the same interferer set at the input port, all FIFO blocking of stream j is accounted for in the FIFO blocking of stream i and must not be accounted here again. ∎

Theorem 4.4.3 The blocking while being in the normal (i.e. low priority) state $B_{i,q}^{dyn}$ of the q-th flit of a stream i can be derived from the threshold value and the event models. It can be denoted as:

$$B_{i,q}^{dyn} \leq \min \left\{ t^{EoT}, \min \left\{ \Delta t \,\middle|\, \sum_{j \in Buf_{P(i)}} \left\{ \eta_j^-(\Delta t) \right\} \geq Q_t \right\} \right\} \tag{4.9}$$

with

$$t^{EoT} = \max \left\{ \delta_{First(p)}^-(q), \hat{B}_{P(First(p)),VC(First(p))}^+(q) \right\}$$

$$+ O_p + \sum_{j \in Tasks(p,k)} R_j^*,$$

where Q_t denotes the threshold value; $\eta_j^-(\Delta t)$ denotes the minimum number of flits of stream j that can arrive in any time interval ΔT; $Buf_{P(i)}$ denotes the set of streams sharing the same buffer with stream i including i; $First(p)$ defines the first task of the chain (i.e. network path); $Tasks(p,k)$ denotes the set of all tasks of path p until the router k under consideration (i.e. the first k routers of the path); O_p denotes the constant de/packetization overhead; R_j^* denotes the worst-case single hop latency assuming no additional blocking through the low priority state; $\delta_{First(p)}^-(q)$ denotes the time the sender needs to inject q flits; and $\hat{B}_1^+(q + b_i - Q_b)$ denotes the overhead induced by backpressure until the q-th flit can be injected to the first router.

Proof. Proven directly. With the definition of the dynamic arbitration, in the worst case, a GT stream has to wait until the threshold value is reached or the EoT flit arrives before it can be selected by the arbiter.

For the former, we need to find the smallest time window in which sufficient events arrive. For this, we need to account for all streams that *must* share the queue. Streams that *might* share the buffer, are not accounted for. All considered streams then arrive with their slowest speed (covered by $\eta_j^-(\Delta t)$). This is covered by the inner *min*-function.

For the latter, $t^m athitEoT$ derives the maximum time until the EoT flit arrives at a router k. It assumes the maximum time needed to inject the EoT flit followed by the worst-case latency to the router under consideration. As the EoT flit is not experiencing any additional delay through the mechanisms, we do not need to account for it when deriving the latency.

As the router prioritized the GT traffic as soon as one of the conditions is fulfilled, we can take the minimum of both values. Covered by the outer *min*-function.   ■

Lemma 4.4.4 — Waiting. The worst-case waiting time $\hat{B}_p^+(q)$ of a port (i.e. buffer) p used by a GT stream denotes the time until the port is ready to receive the q-th flit. For a router it can be bounded by:

$$\hat{B}_p^+(q) = \begin{cases} q \cdot C + \max_{\theta \in \Theta^k} \{A_\theta\}, & \text{if } b_p > Q_b \\ 0, & \text{otherwise} \end{cases}$$

$$\text{with} \quad A_\theta = \sum_{j \in \theta} \left\{ B_i^{out,ep}(\hat{B}_p^+(q) - C_i + \varepsilon, n) + B_{P(j),VC(j)}^{bp}(n) \right\}, \quad (4.10)$$

where b_p denotes the worst-case backlog of the port and k is a bound on the number of packets q flits form ($k = \lceil \frac{q}{n} \rceil$).

Proof. The proof is analogous to the proof for Lemma 3.4.9. The port waiting time (and hence backpressure) can only occur, if the worst-case backlog of the port exceeds the buffer size (i.e. $b_p > Q_b$). In this case, it also exceeds the threshold value for GT and thus the GT stream (e.g. the whole queue) has the highest priority. If backpressure occurs, the port is conservatively assumed to be fully backlogged (i.e. buffer full). Hence, to receive q flits, the port must transmit q flits to any output. For these flits we must account for their transmission time ($q \cdot C$) and the worst-case interference they suffer. For this, the term $\max_{\theta \in \Theta^l} \{A_\theta\}$ obtains the worst-case blocking from each possible mapping of flits to output ports. As the queue has the highest priority, we only need to acccount for output blocking and the waiting time at the next router port.   ■

With all sources of blocking defined, the multiple activation processing time from Equation 4.6 for a single router is fully defined. Based on this we can derive an upper limit for the worst-case single hop latency and end-to-end metrics for the whole network as discussed in Sections 3.4.3 and 3.4.3. Additionally, the worst-case waiting time allows to derive the minimum accepted throughput for a sender at each router port and thus for the whole network according to Equation 3.16 as:

$$\hat{\beta}_p^-(\Delta t) = \min\left\{\lfloor \Delta t - C \rfloor, \max\left\{m \mid \hat{\beta}_p^+(\max(0, m - Q_b)) < \Delta t\right\}\right\} \qquad (4.11)$$

$$with \quad \Delta t \in \mathbb{N}^+.$$

The *max*-function selects the highest number of events that can be accepted during a time interval Δt based on the waiting time. For this, only events that can arrive before Δt can be accepted. As the first Q_b flits can be accepted immediately, we only have to account for the waiting time of $m - Q_b$ flits. Additionally, the port cannot accept more flits than can be physically be transmitted (e.g. one flit every C time units), covered by the first term of the *min*-function. The minimum accepted traffic for a sender then corresponds to the minimum accepted throughput at the first router port on the path of the sender. If this service is equal or higher the requested throughput in the long term, the stream is schedulable (w.r.t. to throughput requirements).

With the known minimum service, we can also derive the backlog b^p based on the minimum service at each port p and hence sender as:

$$b^p = \max\left\{\sum_{i \in Buf_p}\left\{\eta_i^+(\Delta t)\right\} - \hat{\beta}_p^-(\Delta t)\right\}, \qquad (4.12)$$

where Buf_p denotes the set of streams sharing port p. The equation compares the minimum service of the port ($\hat{\beta}_p^-$) with the requested service (η_i^+) of all streams. The difference then shows the maximum number of flits that can have arrived but not be transferred. This equation can be used to derive the size of the buffer at the sender or the bucket size of the shaper, to be able to recover from backlog. For this, the buffer or bucket size must be at least the maximum possible backlog at the sender.

4.4.4 Evaluation

In this section, we evaluate the GT mechanism and compare it against the classic and widely used prioritization scheme [39], and the *backsuction* scheme [58]. We divide the evaluation into two parts. In the first part, we use synthetic workloads to evaluate the basic functioning and certain properties of our mechanism, such as isolation between BE and GT and the influence of different transmission sizes. In the second part, we use memory access and communication traces of general purpose applications, to investigate the performance of the mechanism on realistic workloads. All experiments were carried out with the *OMNeT++* simulation framework and the *HNOCS* library [32] (cf. Section 2.3) using routers with a four-stage pipeline, four virtual channels (VCs), buffers to store 16 flits in each VC, and a packet size of four flits.

As a test scenario, we use a 8×8 mesh NoC with XY-routing, in which we denote the corner node on the north-west as *node (0,0)* and the corner node on the south-east as *node (7,7)* based on their XY coordinates as sketched in Figure 4.12. In the network we have two GT streams, one from node (1,0) to (4,6) requiring 30 % of the link throughput and one from (2,0) to (4,5) requiring 20 %. Hence, both GT streams overlap with a total requirement of 50 % of the link throughput on the shared links. This example is compact enough to be comprehensively displayed but shows all relevant effects.

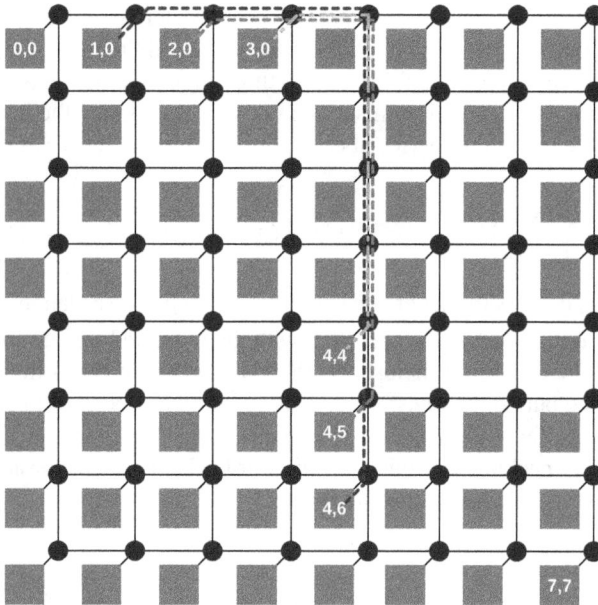

Figure 4.12: *Simple communication scenario for GT.*

Performance

In the first set of experiments we use synthetic workloads, generated based on average link loads. We use a BE stream sending from node (3,0) to node (4,4), hence its whole path is overlapped by the two GT streams. All other nodes

inject BE traffic to random destinations. We then investigate different mechanisms: round-robin (RR), prioritization of GT (SP), backsuction (BS) and
the new approach (FP). As backsuction does not allow to share a virtual
channel between multiple GT streams, the stream from node (1,0) uses *VC0*
and the one from node (2,0) *VC1*, leaving two VCs for BE traffic. For the
other mechanisms, both GT streams share *VC0*. Hence, we additionally
differentiate between the case where we allow BE to use two or three VCs,
denoted respectively as FP2 and FP3 for FP. In the former case, only three
VCs are used, which corresponds to a smaller router design than in the case
for BS that uses four VCs (cf. Table 4.2) for the same set of streams.

Figures 4.13 and 4.14 show the achieved throughput for all approaches
over increasing BE load for the streams *GT1* and *GT2*, where all nodes are
periodically injecting single packets. As can be seen, all QoS mechanisms
provide the GT streams with the required throughput. Only in the round-robin
(RR) case (i.e. no QoS), the *GT1* stream drops below the required throughput
when the BE load increases, which states the RR approach unsuitable for
safety-critical designs.

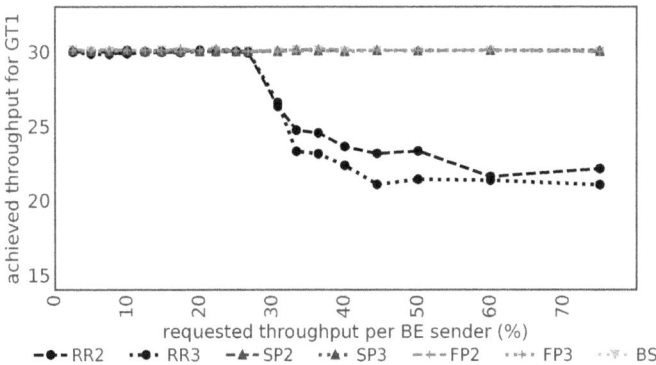

Figure 4.13: *Achieved throughput for GT1 using periodic packets with four flits.*

For the same scenario, Figures 4.15 and 4.16 show the latency for BE
node (3,0) sending to node (4,4) when the GT streams are sending single
packets and bursts of four packets. As expected, the prioritization leads to a
higher latency for BE compared to the RR case. BS and our approach achieve
similar latencies as round robin (RR) for low loads, and hence improve the
latency for BE up to 17 % compared to SP. Along with this, the saturation

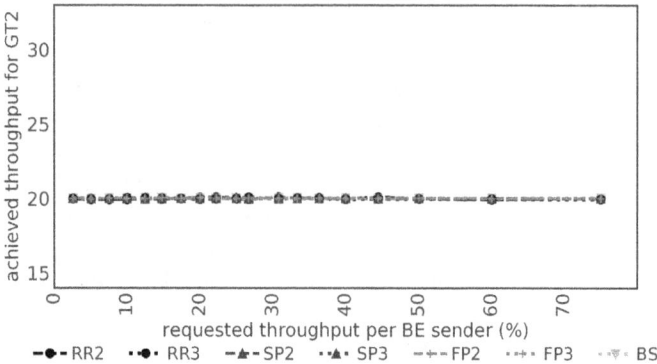

Figure 4.14: *Achieved throughput for GT2 using periodic packets with four flits.*

point of BE, at which the latency for BE traffic goes to infinity, can be shifted to higher workloads enabling a higher system utilization when compared to SP. Here, BS and FP2 achieve a similar performance (while FP2 needs only three VCs). And FP3 (with the same number of VCs as BS) achieves a better performance for BE traffic than BS. Additionally, with an increasing burst size, the latency of BE traffic increases for the simple prioritization. For BS and FP the latency increase for BE traffic is less. Hence, for increased burst sizes, BS and FP can lead to a higher performance improvement for BE traffic.

In the second set of experiments we use benchmark workloads to evaluate the performance of the proposed mechanism. For the experiments we obtained traces from the CHStone benchmark suite [91]. The traces were extracted using the Gem5 simulator and an ARMv7-a core with a 32 kB L1 cache. Each trace contained 100 000 accesses to the network, where each access can be a direct memory access, communication, or a cache access. The compilation was performed using a standard *gcc* compiler (ver. 4.7.3). For a simulation run, we assigned the benchmark under consideration to node (3,0) and then generated several random mappings of the benchmarks to the other nodes with random destinations for their traffic. We selected the destinations such that a traffic stream has to pass at least three routers (i.e. no traffic to direct neighbours).

Figure 4.17 shows the normalized latencies for this scenario. We generated for each listed benchmark 25 different sets of interfering workloads (i.e.

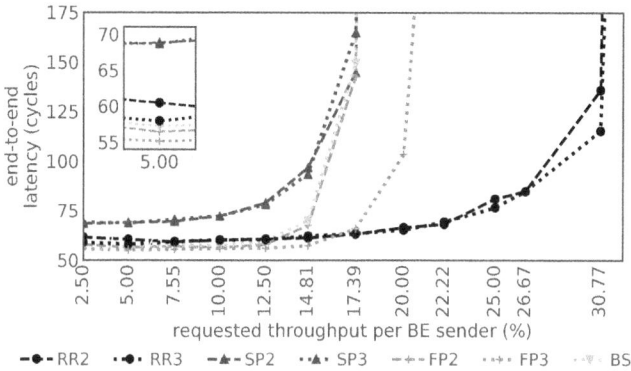

Figure 4.15: *BE Latency periodically sending a single packet.*

random assignment of application to nodes) and a random destination for each sender. We then normalized the latency of the BE sender from node (3,0) to (4,4) to the case of simple prioritization and two VCs for BE (i.e. to SP2). As can be seen, the results comply with the synthetic results, showing that the dynamic prioritization of our approach can improve the BE performance by up to 16 %. Again, our approach leads to similar performance improvements as BS when using less virtual channels or better improvements when using the same number of VCs. Additionally, the experiments show a dependency on the BE traffic patterns. For example, in the case of the *motion* benchmark, there are higher improvements than for the case of *adpcm*.

Synthesis Results

In this section we summarize synthesis results for our approach. We implemented and synthesized a *2×2* NoC on a *Virtex-6 LX760 FPGA* using *Xilinx ISE 14.6* with default optimization settings and no special optimizations for the *VHDL* implementation. The device utilization data were collected from the *Module Level Utilization Summary Report* produced by ISE. As this NoC is not fully connected, each of the four routers has only three input ports fully instantiated. The results for the whole NoC are summarized in Table 4.2. The table compares the used *registers* and *LUTs* for six different implementations. The *RR3* implementation corresponds to a basic round-robin router with four virtual channels (i.e. three VCs for BE and one for GT) and a buffer depth of four packets for each VC. This was extended in *SP3* to provide one

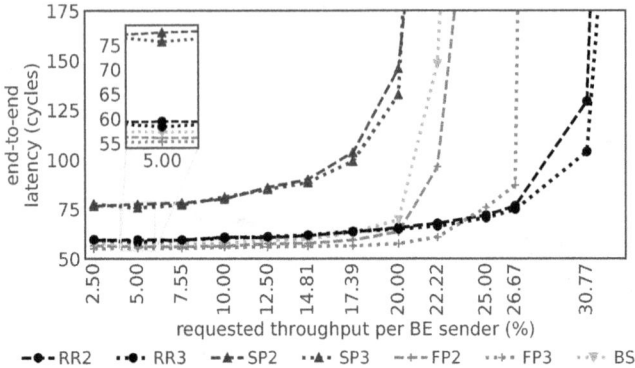

Figure 4.16: *BE Latency periodically sending a burst of four packets.*

prioritized VC for GT (i.e. *VC0*). In the *DP* design the priority of *VC0* can be changed via a configuration flag from the highest to the lowest priority during run-time. This was extended in *BS* to account for the backsuction signal. And finally, the *FP3* and *FP2* implementations denote the new approach (cf. Section 4.4), where the priority of *VC0* is dynamically changed by the router based on the current fill level of the input buffer and the presence of an EoT flit, with respectively four and three VCs.

The synthesis shows that our approach introduces less than 10 % overhead for the used 2×2 NoC compared to a baseline router (RR3) and less than 0.5 % compared to the SP3 design when using the same number of VCs. However, if throughput guarantees are required, the baseline approach (RR3) cannot be used. Here, we need an additional round-robin arbiter at an output port (one for BE and one for GT requests) when going from RR3 to SP3, leading to a higher overhead. When extending SP3 to dynamic priorities, only smaller changes are needed. In comparison to BS, our approach enables to share VCs between different GT streams and hence to lower the number of needed VCs (e.g. FP2) while achieving the same performance improvements (cf. Section 4.4.4) and thus also the overhead.

The achievable clock frequency was 210 MHz for all designs, showing that the extensions did not influence the critical timing path of a router. The frequency was restricted by the minimum achievable period, caused by a *data path delay* of 4.75 ns, consisting of 1.31 ns for logic and 3.44 ns route delay.

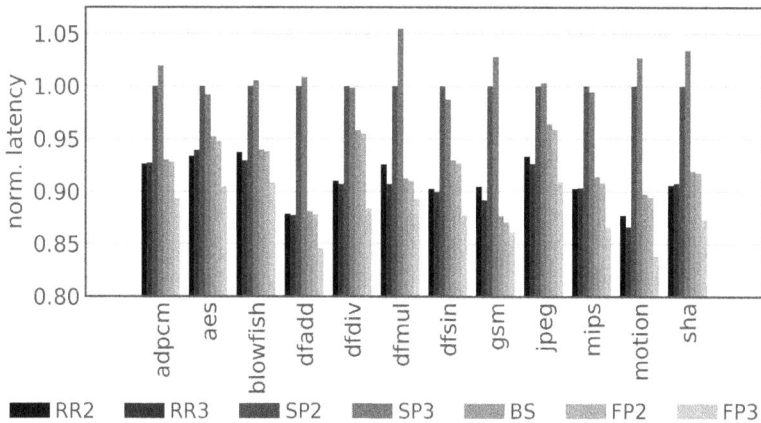

Figure 4.17: *Normalized BE Latency for various benchmarks from node (3,0) to node (4,4).*

Table 4.2: *Synthesis results for 2×2 NoC on Virtex-6 LX760 FPGA.*

Unit	RR3	SP3	DP	BS	FP3	FP2
#Registers	5294	5885	5900	5906	5908	4695
#LUTs	6813	7505	75302	7549	7560	6115
Frequency (MHz)	210	210	210	210	210	210

4.5 NoC Software-control

The presented approaches from Sections 4.3 and 4.4 can improve the performance of the system, especially for BE traffic. However, such hardware based approaches add hardware overhead to the system, which requires to extensively adapt existing NoC designs, and are partly fixed after the design. That is, for example, the number of channels that use these approaches to provide QoS are fixed. Next to the additional control logic overhead of the mechanisms, they might also require a certain minimum buffer size and thus increase the needed buffer space. This is, for example, the case for the GT approach, which benefits from deeper buffers. However, deep buffers require more area and power. And with the fixed design, flexibility and utilization become an issue.

The presented approaches so far work well for a certain traffic class (i.e. either GT or GL) but provide limited or no benefits when applied for the other. Hence, there are basically two different options for a design to use these approaches. First, a static design (e.g. based on a use case) can be derived. Such a design has a fixed number of GL, GT, and BE channels as well as static priority assignments for these. To allow some flexibility, the nodes or network interfaces can decide, which channel will be used by a transmission. However, this still can lead to over- or underutilized channels (e.g. when there is only low or no GT traffic). Hence, the flexibility is limited. Secondly, a runtime configurable design is possible. Here each channel can be configured during runtime to use any of the presented mechanisms. That is, for example, for each channel the GL, GT, and BE mechanisms are implemented and the system can switch between the used mechanism for each channel during run time. While this allows full flexibility, it also has a high hardware overhead.

Another problem, besides the direct hardware and power overhead, is blocking propagation. Sharing of links or virtual channels in a network on chip by several applications may lead to blocking and accumulation of flits of blocked packets in the network. In consequence, these flits may block other streams sharing some links in the network, what is commonly known as a head-of-line (HoL) blocking (cf. Section 4.5.1). This can drastically degrade the performance and increase the power consumption. The typical solution to this problem are deeper buffers that prevent blocked flits to be spread over multiple routers. This is only partly mitigated by the presented HW approaches. While the improved performance for BE reduces the blocking propagation for BE (when the system has low utilization), the blocking for GL/GT increases due to the additional dynamic priority blocking. And, for highly loaded systems, where the approaches might always trigger and thus prioritize GL or GT, BE observes blocking propagation again.

One possible solution to these problems is the use of a software control-layer for resource management of NoC resources [122; 125; 223]. Such NoC resource management (NoC-RM) describes a supervised sharing of network resources (or resources in general), such as the virtual channels (VCs), between applications, using a global arbitration scheme. Network senders synchronize their transmissions through the exchange of special control messages before entering the network. That permits moving blocking from the network switches to the network interfaces or the software stack (e.g. to a higher logical level). In such a way the approach allows to avoid head-of-line blocking in the network and to provide QoS. This can significantly increase the utilization of the interconnect and reduce buffer needs while still allowing flexibility. The approach can be implemented in a centralized as well as a decentralized form as an extension of the network interfaces (NIs) of processing nodes connected to the network [122; 125; 223].

4.5.1 Problem of Blocking Propagation

Modern Networks-on-Chip (NoCs) must not only provide a scalable infrastructure but also deliver high performance at the minimum possible cost and provide flexibility, as discussed in Chapter 1. The efficient co-execution of multiple applications in a system with a NoC requires mechanisms for the allocation of resources to different packets traversing the network. The commonly used wormhole switching offers low-buffer requirements and low average-latencies even at high load [54; 69]. At the same time, wormhole switching results in spreading of packets over routers on their network path, making it sensitive to blocking. To traverse an end-to-end path through a NoC, a packet has to acquire an output port in each router on its way. Because routers are equipped with independent arbiters, this results in an acquisition of a group of resources with local arbitrations. This independent arbitration can lead to problems resulting from a coupled resource allocation. A blocked packet is blocking other packets, even if they are heading towards other destinations due to shared buffers in intermediate router ports. Additionally, the distribution of packets between routers, accelerates the propagation of blocking. This phenomenon is commonly known as *head-of-line* (HoL) blocking [176].

In a result, wormhole switching works well as long as all system modules are capable of receiving and processing packets as they arrive and the congestion time in network routers is short. This is not always possible as many of today's architectures use bandwidth-limited modules, which, due to their cost and complexity, must be intensively shared between applications. Examples are memory modules (e.g. external DRAM) or specialized modules (cache controllers, arithmetic units, special purpose intellectual properties, or SRAM controllers). In the following we will refer to them as *hot-spot (HS) modules*.

If a HS module is not capable to process packets as they arrive the flits of these packets accumulate in buffers of adjacent routers and block other packets causing a domino-like effect. In this manner a saturation tree will form affecting distant parts of the network. For instance, lets consider the example in Figure 4.18. In the example there are eleven IP modules (IP1–11) and a DRAM, which acts as a HS module. It is assumed, that due to the hardware limitations of the system all applications must share the same set of buffers in a NoC, i.e., the same virtual channel. If the DRAM is not capable

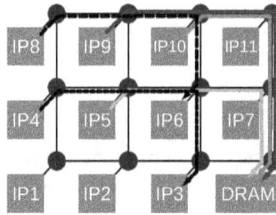

Figure 4.18: *External memory (DRAM) acting as a hot-spot module and causing a saturation tree as well as source unfairness.*

of processing all requests as soon as they arrive, in the worst case, blocked transmissions from IP5 may block the transmission from IP10, which will later block IP9.

Moreover, the propagation of blocking will also affect other communication, which is not directed towards a HS module, i.e., all transmissions sharing the same buffers and links with the senders accessing the HS module. As an example, in Figure 4.18 transmissions from IP8 will be blocked by IP9 and transmissions from IP4 will be blocked by IP5. In consequence, the whole communication in a NoC may be slowed down or even blocked because of a single fully or over-loaded HS module. This results in a drastic decrease of the system performance.

Additionally, the propagation of blocking can cause an unfair distribution of the bandwidth of the HS module between senders. Modules that are close to the HS module will gain access much faster than distant ones. The local arbiters in the network routers usually utilize a round-robin based arbitration between packets from different input ports using the same virtual channel (VC) that compete for the same output port. Hence, they equally divide the bandwidth between the requesters from these inputs. Consequently, in the worst-case the bandwidth available for a particular module decreases exponentially with the number of routers, which its transmission has to pass. In case of a system with HS modules, which are causing frequent blocking, the distance between modules can cause significant differences in transmission latencies, e.g., spatially distant modules may experience long blocking times due to higher probability of blocking. These problems can drastically decrease the NoC performance even in highly over-provisioned networks because even the highest link bandwidth will not prevent blocking propagation resulting from shared buffers. Hence, the problems resulting from hot-spot modules may affect systems with a NoC independently of the link width or available bandwidth as the blocking can originate from the limited processing rate of the particular modules and not only from the NoC itself.

The virtual-channel (VC) flow control allows to mitigate these problems through decoupling of the allocation of buffers from the allocation of bandwidth of a single physical link [54; 69]. However, if there are more applications than available VCs, applications mapped to the same VC will still be able to block each other. Due to buffer limitations (e.g. to save hardware cost) in most architectures multiple applications will share a single (virtual) channel [65; 214; 233]. Hence, other solutions for this problem are necessary.

4.5.2 Control Layer for Resource Management

In real-time systems using NoCs, synchronization between interfering transmissions can help to run the network with significantly less resources and still be able to guarantee temporal properties and safety of a system. Hence, we propose to use a control-layer for resource management of the NoC [122; 125; 223]. The approach provides a global admission control that regulates accesses from different applications to shared resources, such as hot-spot modules and the interconnect. This includes handling the effects of backpressure and head-of-line blocking in routers, where switch arbitration between traffic streams is performed. At the same time, the online control provides flexibility to adapt the system to changes. It allows to simultaneously satisfy performance and safety requirements of modern MPSoCs also in the presence of highly dynamic workloads. Therefore, it allows to overcome the shortcomings of other existing methods requiring modifications of the network routers. The proposed solution is based on a global arbitration and synchronization of accesses to shared resources. It uses the global, current state of the system to online adapt the (local) admission control in NIs and the QoS mechanisms in the used resources (e.g. arbiters in NoC routers).

The base functionality is delivered through a global arbitration layer using key elements of *Software Defined Networking* (SDN) [129] and adapting them for the requirements of real-time NoCs. The scheme decouples the QoS control from traffic arbitration in routers (packet switching and flow control) and separates the NoC in a (virtual) QoS control layer and a data layer as shown in Figure 4.19. This is realized using a protocol-based negotiation between senders, i.e., providing a validation method to check if the currently available NoC resources are sufficient before the application is granted access to the NoC. This also permits adjustments of QoS schemes during the negotiation phase, e.g., different traffic rates depending on the system load. The synchronization can follow through a central scheduling unit or distributed. In both scenarios, the approach can help to relieve the NoC routers from QoS functions, e.g., admission control or maintaining QoS states whether per transmission or aggregate.

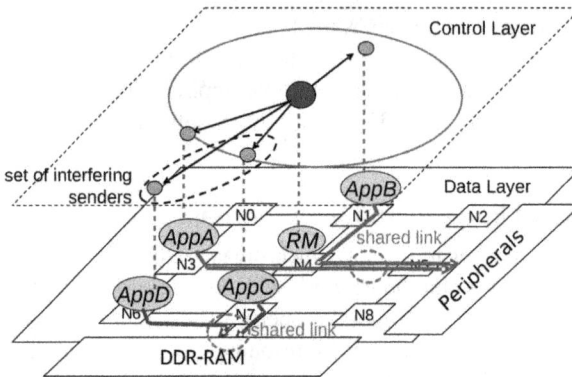

Figure 4.19: *Illustration of the QoS control plane and its operations in the NoC domain.*

Such a decoupling is appealing in several aspects. The major advantage of the proposed approach comes from the fact that the routers can be oblivious of the QoS functions. Therefore, there is no need for custom QoS-oriented NoC extensions, which limit flexibility and can be costly in terms of area and power [81; 174]. The approach allows the deployment of a contract-based QoS provisioning (e.g. round-robin, priority-based arbitration) without introducing complex and hard to maintain schemes, known from hardware arbiters in real-time routers. It can even be applied to commercially available, NoC-based architectures, e.g., Kalray MPPA-256 [42; 65] or Tile64 [214; 233] in real-time domains. Otherwise, these architectures introduce complex interdependences between senders leading to pessimistic worst-case guarantees or lack of formally guaranteed safety [122; 125]. Furthermore, the resource management can be done globally using the knowledge about the current state of the system, e.g., which applications are active, which resources are occupied, and other system and sensor information. Hence, it allows to efficiently incorporate the dynamics of the behaviour of senders and can offer service guarantees by applying temporal-analysis frameworks, such as compositional performance analysis (CPA) [99] or network calculus [134], which allow to capture the dynamics of the system. Additionally, the global resource management, applied in both performance optimized and real-time architectures, allows to improve the utilization of locally and globally shared resources [122; 125; 223]. For instance, the designer may isolate full transmissions, constructed of multiple packets, for ensuring locality of accesses to shared memories or further efficiency improvement [238]. The approach also allows using different resource allocation policies in different regions to improve the overall communication performance of the system [94; 95]. It is also possible to provide feedback about the state of the NoC for effective preemption based on-core schedulers. Consequently, the proposed control layer offers joint benefits integrating features from standard NoCs and real-time NoCs on top of an existing infrastructure. Such approach offers easy implementation, high flexibility, and efficient guarantees even in systems with dynamic workloads.

4.5.3 Principle of Operation of the NoC-RM

To provide service guarantees, an architecture must allow bounding direct and indirect interference between interfering transmissions, i.e., transmissions that overlap in at least one link on their path from source to destination and thus are sharing NoC resources. For doing so, the NoC resource management (NoC-RM) groups applications in *synchronization scenarios*, i.e., sets of senders that may mutually influence their execution times through, for example, concurrent accesses to shared interconnect resources. Ensuring temporal guarantees for synchronization scenarios, i.e., offering a predictable NoC, requires reserving sufficient resources such that the traffic requirements from all senders are met. To achieve this goal, the NoC-RM decouples the admission control and QoS configuration from the system execution and data transport. Hence, the architecture differentiates a *control layer* and a *data layer*. The NoC-RM uses the control layer to provide a validation process performed at runtime before a communication (on the data layer) is established. The validation process checks if current resources are sufficient for the particular transmission and if a reconfiguration is needed and if it is possible. Additionally, it supervises ongoing transmissions and the system state to adapt to (unexpected) changes. Consequently, the main functionality of the NoC-RM encompasses model-based analysis methods, which are used to establish the settings for monitoring and QoS mechanisms. Both layers, the control and data layer, are coupled through a protocol based synchronization, i.e., contracts, allowing resource reservations for senders. For a NoC (or switch-based interconnects in general), the resource reservations can include all routers and links on the end-to-end path. Hence, a reservation can include multiple resources for a single request with selected parameters. Such parameters, for example, describe the QoS or performance requirements, such as the needed throughput on a network path.

The connections for each synchronized sender are negotiated and established dynamically at runtime. The negotiation is based on the global state of the system and specified protocols. The global state of the system is defined by, for example, the number of currently running applications including their requirements w.r.t. shared resources. These factors can change during the runtime depending on the dynamics of the system or applications, as well as the physical environment (e.g. situation on the road for driver assistance systems). Figure 4.20 shows an exemplary design for a model based adaptation. In the example, the control layer comprises multiple *analysis engines* responsible for different aspects of resource allocation and admission control. The analysis can be done for optimizing different system goals (e.g. safety, security, or performance) and provides an initial input to the *model exploration* module. Moreover, the analysis engines can assign resources according to pre-defined and static allocation schemes (e.g. time-division multiplexing, TDM) as well as dynamic and work conserving schedulers (e.g. round-robin, static or dynamic priority based policies). The *model exploration* component is optional and used to perform design-space exploration within the system model to find a feasible solution whenever the requested contract cannot be established as well as to adjust system to unexpected events, e.g., robustness against faults of the components. Consequently, it introduces adaptivity and self awareness.

As the main goal of this work is to introduce a predictable and runtime-adaptable NoC, the focus lies on predictable resource allocation schemes that are defined during the system design phase and controlled by the resource management. Consequently, the main goals of the arbitration enforced by the NoC-RM in this work can be defined as follows:

- avoiding contention in (NoC) buffers;
- dividing bandwidth between interfering senders according to their requirements;
- adjusting the settings during runtime to cope with dynamics in the behaviour.

The NoC-RM opens multiple implementation possibilities. A comprehensive overview of different protocols is presented in [122; 125; 223]. In this work we will focus on the low level (e.g. hardware) architecture needed to efficiently support the resource management.

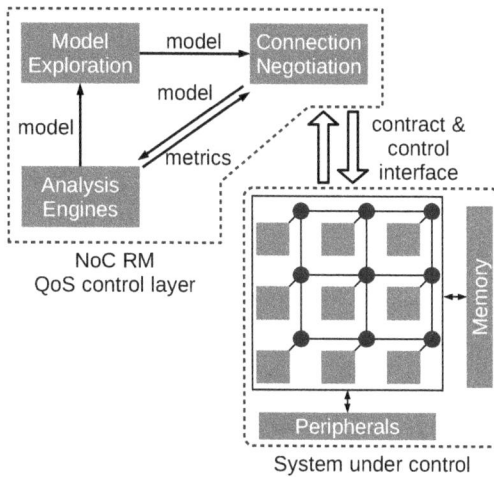

Figure 4.20: *Model domain architecture for adaptive arbitration in a NoC.*

4.5.4 Synchronization in NoC-RM

The synchronization process of the global resource management can be divided into the following phases: (1) *initialization/requesting*, (2) *reservation/negotiation*, (3) *usage*, (4) *release*. Figure 4.21 provides a generalized version of the synchronization workflow. The actual steps and their details may vary according the selected protocol and implementation [125; 223].

To start a communication an initialization procedure (phases 1 and 2) must be performed. First, a sender, willing to communicate with a particular receiver, tries to access the communication path and thus certain resources in the network. The NoC-RM then checks if the access needs a synchronization based on the current state of the system. For this, the NoC-RM uses the state of the system and available communication scenarios The communication scenario is defined through the communication between all different senders and receivers that may happen simultaneously in the network. Based on these scenarios the NoC-RM decides if a new access can be granted or if the request must be serialized and thus has to wait for another state in the network. After receiving a confirmation, the sender can access the network of resource. The fourth communication phase happens when a sender finishes using a communication path. The end of the communication is recognized by the NoC-RM (or announced by the sender). Based on this information,

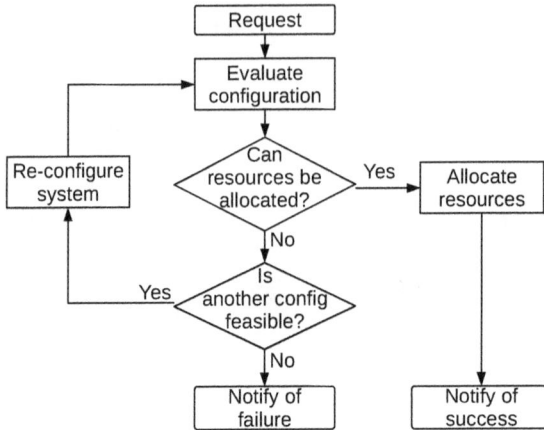

Figure 4.21: *Generalized workflow of the NoC-RM.*

the NoC-RM releases the resources and enables the changes to another (communication) scenario. All four stages of the resource management mechanism can be transparent to the sender. Because of that, no modifications of the applications running in the system are necessary.

With the NoC-RM, a decision is made, whether an access can be permitted or not before (data) packets are sent to the network. This allows delaying packets, which could not be served, before they enter the NoC. As the blocking happens outside the NoC infrastructure, buffers in the routers are free and thus blocking propagation can be avoided. In a result, other ongoing transmissions in the NoC are unaffected and the overall utilization increases drastically during the congestion periods. To illustrate that lets consider once again the example depicted in Figure 4.18. In a worst-case scenario, if IP7 accessing the DRAM HS module will get blocked due to saturation then it will also block IP8, although IP8 is communicating with a different module, assuming a standard NoC with wormhole switching. If resource managements is applied the transmission from IP7 is blocked and thus will be delayed by the NoC-RM before the packets will physically enter the interconnect. This permits IP8 to use the released resources and possibly to conduct multiple transmissions. Figure 4.22 shows the effect of this.

However, such a resource management also introduces a set of new challenges that must be considered during the system design phase. The protocol based synchronization is done at the cost of an additional temporal

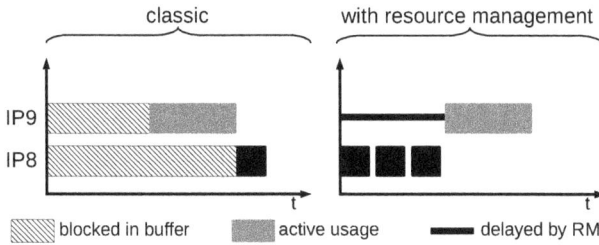

Figure 4.22: *Effects of local arbitration (left) and decentralized resource management (right) on temporal behaviour of IP7 and IP8 from scenario in Figure 4.18.*

overhead. Consequently, the scalability of the architecture and its ability to manage a high number of requesters and many complex scenarios plays a critical role for future industrial deployments. To efficiently conduct the different steps, the underlying (hardware) architecture needs to provide support for the control layer of the NoC-RM (cf. Section 5).

4.5.5 High-level Architecture of the NoC-RM

The dynamic resource management requires a safe modification of resource allocation at runtime for adjusting the QoS settings of the NoC to the changing, global state of the system. This requires the consideration of the local state of the core (e.g. the number and execution profiles of currently active applications defining the requirements w.r.t. the NoC) as well as to the global state of the NoC (e.g. the number and profiles of active interfering senders defining possible interference from interconnect). Although the evaluation of the former *on-core* factors can be done locally on the same node, information about the latter *off-core* factors requires further, global synchronization done by the NoC-RM.

For the NoC resource management, each network node is controlled by a supervisor (*client*) monitoring and controlling all transmissions from running senders. These are used to control the network interface (e.g. to configure rate limiters and to delay a network access until it is allowed by the NoC-RM) and to negotiate the network access. For the negotiation, we distinguish a centralized and a distributed approach for the control layer of the NoC-RM. In the distributed approach clients exchange control messages with each other containing information about the desired resource usages and requirements. In the centralized approach, clients send a request to a central scheduling unit (the *resource manager*).

Centralized Control Layer

Figure 4.23 presents a high level abstraction of a centralized implementation of the control layer. It uses a *resource manager* (RM), which is responsible for decision-making, and clients at the networks nodes, which are responsible for negotiation resource accesses with the RM. Hence, the clients and the RM provide connection oriented network services to provide a predictable and safe behaviour. To use a connection-oriented network service, the communication participant (e.g. IP) must first establish a connection, as discussed in the previous section, then use the resources, and release it at the end.

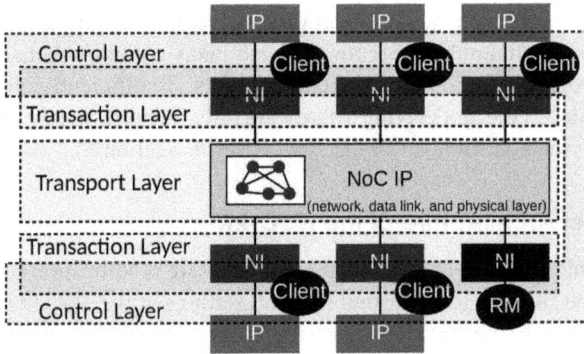

Figure 4.23: *Modules and architecture of the centralized resource management.*

For this, we introduce *clients* at each node monitoring and controlling all transmissions from running senders. Clients are responsible for:

- establishing connections with the RM (issuing correct messages to the RM for appropriate actions of senders and processing node);
- preventing non-authorized accesses to the NoC;
- adjusting local admission setting in NI, e.g., rate settings for rate limiters or MMU/MPU address translation tables, based on the configuration messages from RM;
- preventing too frequent accesses;
- releasing the NoC resources (i.e. inform the RM whenever an application terminates); and

- preventing unbounded NoC accesses, e.g., release the NoC resources (inform the RM whenever a connection takes to long or end too long connections).

The workflow of the mechanisms involves four stages as mentioned above: (1) *initialization/requesting* a transmission, (2) *reservation/negotiation* of resource/network access, (3) the *transmission* itself, and (4) *releasing* the resources.

For the initialization procedure (stages 1 and 2) the client traps the access of the sender to the network to prevent unauthorized transmissions. The client has information about the current state of all connections allowed from the particular node. Depending on the current state of the resources a request to the resource manager is generated or the access is directly forwarded to the resource (i.e. when the access was already granted and is still allowed). During the second stage, the resource manager processes the request from the client. This includes checking the global state of the network and available communication scenarios. The communication scenario is defined through the communication between all different senders and receivers that may happen simultaneously in the network. Based on these scenarios the RM decides if a new access can be granted or if the request must be serialized and thus has to wait for another state in the network. Later the resource manager sends a response to the client with information about the acceptance or rejection of the request. After receiving the confirmation, the client forwards the access, transparently from the sender, to the network. Starting from that moment the sender may use the resource. After granting the access, the client monitors the transmission from the sender. The fourth communication phase happens when a sender finishes using a communication path. The client recognizes the end of the transmission and informs the RM that the communication is over. After receiving the notification, the RM has to release the resources and to enable the change to another (communication) scenario. All four stages of the resource management mechanism can be transparent to the sender. Because of that, no modifications of the applications running in the system are necessary.

The client modules can be implemented as hardware extensions of the NI for high performance or as software modules running on the IP core (however also in this case some extension of the NI could be necessary). Similarly, the RM can be realized fully in hardware, i.e., as an independent HW IP connected to the NoC, or as a software IP running on one of the nodes. The main advantages of the software realization is flexibility. The complexity of both units depends on the selected synchronization protocol. More details on this can be found in [122; 125].

Separating the clients and RMs from the software running on the processing nodes is especially important in safety-critical systems where the NoC is treated as a shared resource. Recall, that in such systems it is necessary to either certify the whole system (including all applications) to the highest relevant safety level or decouple the resource arbitration from senders to provide sufficient independence (cf. Section 1.2) [7; 15]. As ensuring adherence to standards is a costly and demanding process, the latter is the preferred solution for most systems. This can be achieved through the clients and RM, which can be designed (and certified) independently to the highest relevant safety level.

Distributed Control Layer

Figure 4.24 presents a high level abstraction of a distributed implementation of the control layer. In the distributed approach, there is no central resource manager. It uses clients at the networks nodes, which are responsible for negotiation resource accesses in a distributed manner. Hence, these clients are a more complex component compared to the ones of the centralized approach. That way, they provide connection oriented network services to provide a predictable and safe behaviour. To use a connection-oriented network service, the communication participant (e.g. IP) must first establish a connection, as discussed in the previous section, then use the resources and release it at the end.

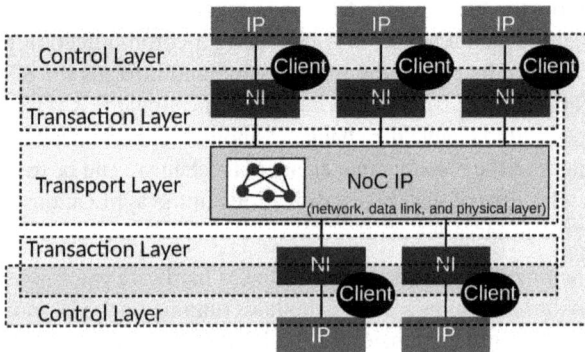

Figure 4.24: Modules and architecture of the distributed resource management.

The synchronization is done through the clients, controlling the state of the transmissions. In a system utilizing the decentralized resource management there is one client per network node. An offline analysis defines all possible communication scenarios. Basing on these scenarios it is necessary to decide which resources can be used without supervision and which require synchronization of the accesses. A synchronization scenario then defines all the applications sharing the same resources which are not allowed to access it simultaneously or appropriate rate limitation when accessing them simultaneously. Each of these applications will be later classified as a sender and supervised by a client.

The workflow is similar to the centralized approach and also involves four stages: 1) *initialization/requesting* a transmission, (2) *reservation/negotiation* of resource/network access, (3) the *transmission* itself, and (4) *releasing* the resources. However, there is a difference in the second stage now. Similar as before, a sender tries to access the network (or a resource) to start a communication. This access is trapped by the client, which checks if the access can be allowed directly or if a negotiation or re-configuration is needed. If a negotiation is needed, the client sends requests to the all corresponding clients for the selected scenario. The scenario is defined by the requested resources and includes a list of involved clients. During the second stage, the clients use a distributed decision-making protocol, to decide if the request can be granted and if a re-configuration is needed. If the access is granted, the transmission can start using the decided parameters. When a sender finishes using the resources, another negotiation is initiated to inform the other clients, that the resources are released. Again, all four stages of the resource management mechanism can be transparent to the sender.

4.6 Summary

QoS support plays a crucial role for safety-critical real-time systems. Without QoS platforms cannot provide guarantees on the timing behaviour or only very pessimistic estimations rendering many design unfeasible. At the same time, the efficiency (e.g. high performance, low area and power overhead) and flexibility are of high importance to reduce the costs of the overall system.

Hardware based QoS mechanisms can only solve some of these challenges. While a HW mechanism can be very efficient for a certain use case (e.g. when the design is optimized for the use case), its flexibility is limited. As soon as the use case exhibits dynamics or the system should be used for a different scenario (e.g. system update, new functions added, different use case), the benefits are mitigated. That is, the system cannot be fully utilized and might waste performance or power.

Software based mechanisms, on the other hand, can provide efficient QoS and flexibility at the same time. Additionally, they can be applied to existing systems without the need for extensive adaptations of existing hardware components, further reducing (development) costs. The NoC-RM provides a QoS abstraction of the underlying NoC (data plane) allowing a path oriented approach in which the per-hop behaviours of routers and the end-to-end properties of communication can be unified. This allows safe and efficient resource reservations but requires knowledge about the global state of the system, i.e., the number of simultaneously running senders, their current state and the QoS requirements, which may change during runtime. Therefore, for establishing and adaptively managing connections, clients are used to negotiate reservations and resource accesses. This is done through the exchange of messages between nodes with interfering senders or with a centralized control unit. This communication is forming a synchronization protocol allowing to propagate information about the state of a connection as well as the current QoS requirements for a particular sender. The latter parameter can change dynamically during runtime and a contract based negotiation allows to safely adjust the QoS of the platform.

The main advantage of the NoC resource management is that it makes the QoS functions in the routers oblivious for achieving real-time and safety guarantees. The admission control and adaptive management of the NoC state is fully controlled by the introduced control layer making the system potentially more efficient. This allows to adaptively optimize the arbitration to the changing global state of the system for accommodating arriving workloads as well as reacting to possible errors and changes in the environmental conditions.

The NoC-RM can apply different isolation techniques to achieve the aforementioned goals [122; 125; 223]. This flexibility allows the designer to extend the capabilities of a selected NoC-based platform without the need for modifications of the routers. For instance, the introduced solution permits the implementation of a TDM-based arbitration on top of a NoC with performance oriented arbiters in routers such as round-robin based iSLIP arbiters. Consequently, the NoC-RM permits increasing the spectrum of system applications, e.g., adjusting it for providing guarantees in real-time domains, without need for re-design or costly hardware extensions.

However, the benefits of the NoC-RM strongly depend on the (hardware) platform support for the control layer. Hence, to fully utilize the performance and flexibility, the control layer also requires some hardware extensions in the system. Some of the needed hardware extensions might already be present in an existing architecture. Still, the adaptability and reuse-ability make such a software based QoS mechanisms a reasonable choice. Hence, the next chapter derives an exemplary architecture supporting the control layer to fully utilize the benefits of the presented resource management.

5. NoC Architecture Supporting a Control-layer

Today's and future safety-critical real-time systems require performance isolation and adaptability at the same time. The network-on-chip (NoC) resource management (NoC-RM) with its control layer can reach similar isolation properties as complex QoS-aware router designs (cf. Section 4.5). At the same time, it offers an adaptability, while requiring less hardware, allowing the use of COTS networks, and achieving better performance in the average case. This chapter presents an architecture efficiently supporting the NoC-RM to improve its benefits and simplify the data layer network architecture (e.g. hardware complexity).

The chapter is partially based on the work published in [216; 223].

5.1 Introduction

Today's and future safety-critical real-time systems and thus their interconnects must face new challenges (cf. Chapter 1). NoCs must efficiently host new sets of highly dynamic workloads and the behaviour of the system may be influenced by external factors. Tasks, for example, may modify their transmission profiles at run-time depending on the arriving sensor data. Moreover, sets of applications may be initiated dynamically during the runtime of the system, introducing modes with changing traffic patterns and mapping or workload profiles, e.g., convolution of a neural network used for decision-making. Similarly, sensor fusion and data crunching bring up data-dependent execution, resulting in a highly dynamic task behaviour and large jitters.

While classic hardware based quality of service (QoS) mechanisms (e.g. complex routers) can provide the needed performance isolation (cf. Sections 4.3 and 4.4), they cannot efficiently handle such dynamic workloads or adapt to a changing system behaviour. Software control can reach similar isolation properties as complex QoS-aware router designs (cf. Section 4.5). Additionally, a software approach allows an online adaptable design, the use of simple commercial off-the-shelf (COTS) routers, and can even achieve better average case performance [125; 223]. Hence, an optimised architecture supporting a control layer for the NoC-RM for providing QoS and improving the performance of the control layer is needed for its full benefits.

Using the NoC-RM and a control layer has multiple impacts on the architecture design. First, the architecture needs to provide support for the control layer to increase its benefits. The latency of control messages directly influences the synchronisation overhead [120; 122; 125; 223]. Hence, an architecture needs to minimize it. Secondly, if a control layer is available, the hardware for the data layer can be simplified. That is, we can use less complex QoS mechanisms, less virtual channels, and smaller buffers in the routers. This can drastically reduce the hardware costs of the system. Hence, the additional goals of an architecture supporting the control layer are: decreasing the latency of control messages and improving the performance to hardware cost ratio. At the same time, it should not challenge the safety and performance of the system or control layer.

Besides the performance and hardware overhead, a control layer can also influence other parts of a NoC architecture. The control layer can, for example, also be applied for power or reliability management. That is, the control layer can be used to adapt the voltage and frequency settings of the data layer components, power off/on the data layer components completely, or configure redundant/alternate routes in case of errors [123].

This chapter presents an architecture supporting the NoC resource management, as described in Section 4.5, to simplify the network architecture (e.g. hardware complexity) and providing support for the control layer to increase its efficiency. The architecture allows an efficient integration of the control layer and uses it so save cost (i.e. allow the use of simple routers and avoid head of line blocking in the network). Thus, it reduces the complexity of the NoC (routers) and the negative effects of the control messages as well as improves the performance of the control layer. As NoCs are often application specific and several parameters, such as topology, routing, packet (flit, phit), and link size, may vary, the presented architecture is only an example showing the general idea of supporting the control layer. The control layer can be applied to any data layer architecture and thus help to simplify the data layer (e.g. move QoS support from routers to the network boarder and reduce the needed buffer space in the network). Following the definition of [132], the proposed NoC resource management influences mainly the transaction or transport layer, others (i.e. physical, data link, network layer) are technology and use case dependent and, thus, may vary between different designs.

The remainder of this chapter is organized as follows. Section 5.2 provides an overview on the general design requirements. And Section 5.3 presents the proposed architecture. Chapter 6 will then evaluate the design.

5.2 Requirements

For the design of an architecture utilizing a NoC-RM, as introduced in Section 4.5, several aspects and requirements play a role. First, the general requirements of safety-critical real-time systems and their different traffic classes must be accounted. These limit the set of possible implementations for the control layer. And secondly, the requirements of the NoC-RM directly influence the design.

The general requirements of safety-critical real-time systems were already introduced in Section 1.4. In a nut shell, embedded systems in this domain host applications with different traffic characteristics and requirements. We can differentiate at least three different traffic classes: *guaranteed latency (GL)*, *guaranteed throughput* (GT), and *best-effort* (BE). GL requires a limited upper bound for the latency (e.g. to stay below a deadline) and possibly also a comparatively low latency. This limit can be on flit or packet level as well as for bursts of packets. GT, on the other hand, does not have strict latency requirements for single packets, but requires the system to guarantee a minimum accepted throughput. And BE does not have any requirements that must be guaranteed by the system. However, the performance of BE can be of interest to increase the user experience or provide additional, non safety-critical functions. Hence, as long as the requirements of GL and GT can be satisfied, the system should optimize the performance of BE traffic [217].

Next to this, we have the requirements and properties of the NoC-RM. The actual requirements depend on the used protocol and variant of the NoC-RM [122; 125; 223]. But a few requirements are common between them. Hence, we will focus on the general requirements and design considerations for the underlying network. The requirements of the NoC-RM can be divided into the requirements resulting from the *control protocol* and needed modules, such as the *client interfaces* and the *control unit RM* for the centralized approach or the *advanced client interfaces* (including the functionality of the RM) for the distributed approach. Independently of the selected protocol variant the client interfaces must fulfil the following functional requirements [122; 125; 223]:

- intercepting and distinguishing between different transmissions;
- generating and sending of request messages;
- processing of acknowledge messages;
- processing of configuration messages and, for example, changing the rate control (including delaying the access to the resource);
- detecting the end of a transmission and releasing resources (e.g. generating and sending of release messages).

These actions of the client interfaces can be implemented transparently to the running applications, ensuring compatibility with existing legacy software, or as an additional resource (hardware or software library) that must be actively used by the running software (e.g. application or OS). The decision on this will directly influence the low-level implementation of the clients. Kostrzewa [122] presents details on different implementation variants. From the view point of the network, these possible client implementations do not have an influence on the architecture of the network and only influence the details of the network interface (NI) implementation.

Additionally, it is important to decide the granularity level of the synchronisation, i.e. the amount of data on which the client interfaces negotiate transmissions and the information included in a synchronisation scenario. Client interfaces may conduct the synchronisation depending on various factors, including for instance [122; 223]:

- each initiated transmission from the processing node;
- transmissions using a particular set of resources (VC or path);
- transmissions targeting a particular receiver;
- transmissions initialized by a particular task/application or module within the processing node, e.g., DMA engine or a specific task running within an OS;
- using monitoring information, e.g., continuous synchronisation depending on the frequency of initiated transmissions.
- each based on:
  - single packets;
  - full transmissions (i.e. multiple/many packets); or
  - modes of the application.

The decision on these parameters influences the implementation details of the client interfaces (such as the needed local information stored at the client) and the needed information in the control messages and thus the size of the control messages and the implementation of the RM (or other receiving nodes). Hence, these can influence the hardware and temporal overhead of the control layer and its effect on the network architecture. The design must ensure that the processing time and the consumed chip area stay within reasonable bounds.

Besides the implementation decisions, the RM control unit (centralized approach) or clients (decentralized approach) must fulfil certain functional requirements:

- receiving and qualifying control messages (type and sender);
- distinguishing between different synchronization scenarios;
- switching between scenarios according to the predefined set of rules (scheduling method);
- ensuring safe transition between system states (preventing sporadic overloads); and
- generating control messages.

Next to these general requirements, the protocol and clients or RM can be enhanced with more advanced mechanisms and features. This will extend the set of requirements and also may increase the hardware and temporal overhead. The hardware overhead additionally depends directly on the complexity of the introduced arbitration (e.g. dynamic switching between off-line defined scenarios or full online resource allocation and reconfiguration), verification methods as well as availability of existing infrastructure for a particular configuration of a chip. In many systems this cost can be amortized by the reusability of existing system components, e.g., monitoring infrastructure. Kostrzewa [122] investigated this trade-off between design complexity and functionality. However, as it does not directly influence the under-laying network architecture, we will not handle it in this work.

Breaking all the requirements down to the level of the network-on-chip, we can formulate three main design goals of the control layer for the support of the NoC-RM:

1. low latency of control messages, which will reduce the temporal overhead of the control layer;

2. low interference of the control messages on existing traffic to minimize additional interference and blocking in the network induced by the control messages itself; and

3. low hardware overhead (of the overall design), especially regarding the network routers.

The latter is important, as NoCs are already complex systems with a multitude of design parameters and already consume a significant proportion of the die area and system power [40; 41]. For example, the network of Intel's 80-Core Teraflops Research Chip consumes 17 % of die area and 28 % of system power [101; 225]. And the network buffers of the Intel's Teraflops Research Chip consume 22 % of the communication power [225]. The iMesh NoC of the TILE64 many-core system dedicates 60 % of the NoC's die area to buffering [233]. Similarly, in the TRIPS chip, the input buffers occupy 75 % of the router's die area [85]. Consequently, reducing the needed buffers or the needed additional overhead can help to reduce the area and power requirements of NoCs.

5.3 Architecture Details

In this section we derive an architecture of the control layer for supporting the NoC-RM introduced in Section 4.5. We focus on the support for the NoC-RM to provide QoS. As the NoC-RM allows different implementation and protocol variants, we show how the general properties influence the design decisions of the NoC and only present an exemplary architecture. Hence, for certain use cases or protocols this architecture must be adapted. We will differentiate a *data layer* and a *control layer* and highlight the general interaction between the layers. Additionally, we outline concepts extending the control layer beyond its admission control capabilities. Figure 5.1 shows a high level representation of the architecture. It consists of a transaction layer, a transport layer, and the interconnected IP cores. The transaction layer handles the end-to-end transmission of messages as well as network admission control. Its main components are the network interfaces, connecting the IPs/nodes to the network. The transport layer is responsible for moving the messages from one IP to another. These messages can be on the data or control layer. And the IPs represent the processing units, memories, and other on-chip resources as well as off-chip interfaces.

Using the NoC-RM, we can distinguish at least four different traffic classes in the network: *control traffic (CT), guaranteed latency (GL), guaranteed throughput (GT)*, and *best-effort (BE)*. For GL and GT, the network

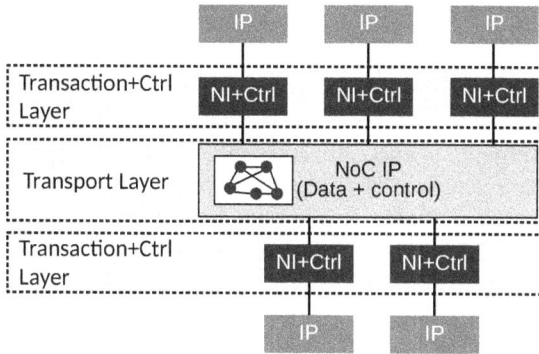

Figure 5.1: *Logical layers of the proposed NoC architecture.*

must provide upper bounds on the latency and lower bounds on the minimum throughput available for the streams. Control traffic (CT) is similar to GL traffic, as an upper bound on the latency is needed to provide a predictable NoC-RM. However, as the performance of the NoC-RM directly influences the performance of the (controlled) system, CT must also achieve a low latency (while for GL traffic it is sufficient to reach a destination by the deadline and not as fast as possible [201; 217]).

To provide sufficient isolation between the different classes (e.g. avoid head of line blocking), several network channels can be used, i.e., one channel per class. If head of line blocking can be avoided, it is also possible to map certain classes on the same channel. When four different channels are used, these can be implemented as virtual or physical channels. Hence, the four traffic classes can be implemented on four different virtual channels, four physical channels (e.g. independent networks), or any hybrid solution, as for example two physical channels with virtual channels on each. Additionally, it is possible to provide multiple channels for a class, as for example, four independent networks (for each traffic class one) with virtual channels on each.

5.3.1 Virtual or Physical Control Layer

The latency of control messages has direct influence on the performance of the NoC-RM. A higher latency will increase the synchronisation overhead and thus can limit the performance benefits or even render the synchronisation as infeasible for some safety-critical applications. There are two different possibilities for implementing the control layer: using the existing data layer or implementing a dedicated control interconnect.

Using the existing data layer is straightforward and often needs no adaptations of the existing interconnect. The only requirement is that the interconnect can provide latency guarantees for the control messages. While a performance oriented architecture, e.g., with round-robin arbitration, is possible, it decreases the performance and predictability of the control layer. Hence, the control layer should be prioritized, e.g., using a high priority virtual channel. This approach can reduce the hardware overhead of the control layer and avoid any modifications to the interconnect. However, when there is no free virtual channel (and the existing traffic of the individual VCs should not be merged), an additional virtual channel must be implemented, which in turn requires additional buffer space in the network routers. Additionally, as the control and data messages share the same hardware, the control messages interfere with the data messages. That is, the control layer induces an additional (high priority) interference to the network, possibly increasing the latencies of the data layer (cf. Chapter 3). To reduce the overhead in this case, the control layer should use the shortest message size (e.g. single flits) to minimize the induced blocking. Simulation and analysis can be used to verify if the introduced benefits of the control layer will outweigh the overhead.

To avoid the additional interference, a dedicated control NoC can be used. Similar approaches are already used by existing architectures to avoid interference between different kinds of traffic, as for example in the Tile64 [214; 233] or MPPA [42; 65]. Such an approach can improve the performance and energy-delay product of a system, especially for regular traffic where collisions are less frequent [26; 46; 237]. As the control traffic follows a well-defined protocol and, for the case of the centralized approached, has a limited set of possible destinations for the packets, it is regular and allows to deploy shorter queues in the routers [122]. In this approach, the control interconnect can be an additional network-on-chip, a bus, dedicated signal wires, or any other interconnect.

Using a dedicated control interconnect allows to use a simple architecture for the data layer and to reduce the depths of the buffers (e.g. as there is less interference compared to the approach with high priority control messages). Hence, the additional hardware overhead of a dedicated control interconnect compared to an additional virtual channel is small. Additionally, as the control layer sends fewer data, the interconnect can be optimized to the control protocol (e.g. smaller message sizes, lower frequency of routers). Hence, a dedicated physical channel might induce less overhead compared to an additional virtual channel that uses the same packet and flit format as the data layer.

From the design perspective, a dedicated control interconnect also offers some other advantages. With a dedicated interconnect, the control layer can easily be re-used in different designs, without costly adaptations. This allows to re-use a well-tested architecture without the need for costly verifications in each new platform. The additional control interconnect can then be used to monitor and supervise an existing data layer and ease the verification of it. This allows the control layer to gather monitoring information of the data layer during runtime, enabling an online supervision and control of the communication architecture.

Overall, an additional physical control interconnect seems a reasonable choice. This leads to an architecture as shown in Figure 5.2, where we use a dedicated NoC for each interconnect. In the example, the control layer is a small optimized NoC, e.g., without virtual channels and only single flit buffers (cf. Section 5.3.3). The control routers can have a monitoring and control module directly connected to the routers of the data layer. The data layer uses virtual channels to distinguish between different traffic classes and a priority based arbitration between the VCs (cf. Section 5.3.2). As the flits inside the data layer typically experience less contention due to the NoC-RM, the data layer uses only small buffers to hide link/control flow latency.

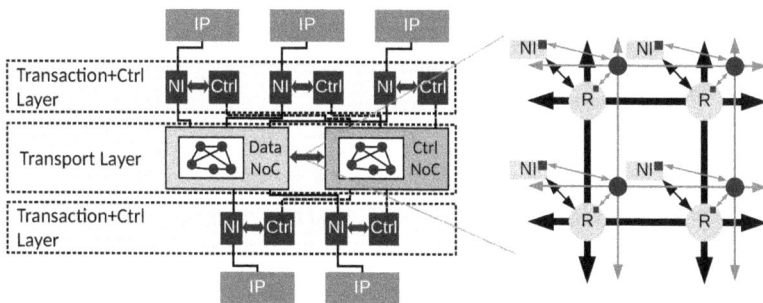

Figure 5.2: *High level NoC architecture using a dedicated control layer.*

5.3.2 Data Transport Layer

The data layer of the architecture is based on the IDAMC architecture and using a priority based arbitration (cf. Section 2.2) [128; 186; 206; 216]. We use a 16×16 2D mesh topology as the baseline NoC architecture. The focus in this chapter is on QoS and the control layer. Other aspects, such as the NoC size, packet format of the data layer, or the routing and its encoding, are just examples and have no effect on the general working of the QoS provisioning of the control layer.

The network routers use wormhole switching, input buffering of incoming data, and implement up to four *virtual channels (VC)*, i.e., separate buffers that share the same link. Figure 5.3 shows an example for the router architecture with n ports. The number of virtual channels and ports are based on the decisions for the size of the header and the used encoding of the routing. If the number of channels or ports is reduced, the header and possibly the link width can also be reduced.

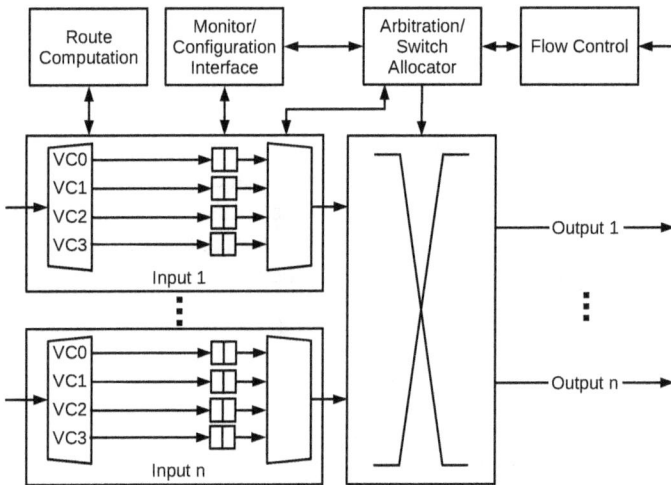

Figure 5.3: *Proposed router architecture of the data layer*

Switching and Flow Control

The NoC transfers data in packets and uses wormhole switching. Duato et al. [69] showed that wormhole switching outperforms *virtual cut through* (VCT) and *store and forward* (S&F) switching. VCT and wormhole switching can achieve better latencies than S&F switching. At the same time, virtual-channel based wormhole-switching has similar buffer capacity than VCT. However, VC-based wormhole switching can achieve higher throughput. While VCT with virtual channels can achieve similar throughput, the buffer requirements are typically higher. From the design perspective, an architecture with a low number of VCs is preferable to save cost and power, whereas a higher number of VCs can improve performance. Pande et al. [166] reported that the optimum number of VCs is four, as a further increase of the number shows only marginal performance improvements while increasing power and area overhead. Hence, the data layer uses four virtual channels in the basic configuration, while other designs are possible. While this is one channel more than needed for the three traffic classes of the data layer, the control layer (or designer) can use the additional channel for traffic distribution or a more fine granular differentiation of traffic.

The VCs are assigned to a priority level and a priority based scheduling is used between different VCs, whereas round-robin is used between VCs with the same priority [206]. That is, the data layer supports four priority levels. QoS is based on reservation of virtual channels for individual traffic streams. There are two possibilities. First, during design time, sets of VC buffers are reserved for certain guaranteed service (GS) streams. Sharing of VCs of different streams or traffic classes can be allowed, when, for example, source rate limiting limits the interference. Second, the NoC-RM can assign streams to VCs during runtime, based on the current state of the system (cf. Section 4.5). The network interfaces (NIs) then map the traffic streams to the selected VCs. Hence, the traffic is not allowed to change the VC on its path through the network. Doing this, the VC allocation step inside the routers can be avoided. This enables to implement virtual networks similar to the work of Heisswolf et al. [95]. For both cases, the control layer can be used to decide which traffic streams are allowed to use the NoC concurrently, at what time, and the allowed traffic rate.

The size of the input buffers is two flits. However, there is the possibility to use different buffer sizes for the VCs [206]. This buffer size is smaller than for a typical NoC switch as shown in Section 2.1.6. As we use the NoC-RM to prevent complex blocking situations and blocking propagation, we do not need deep buffers to mitigate HoL blocking as would be needed for VCT. To avoid buffer overflow between routers, the network uses a credit based flow control scheme allowing backpressure to occur. For this, the flow control keeps track of available buffer space at the down stream routers using credits. Each time a flit is sent downstream, the credit counter for this particular port and virtual channel is decreased. When the switch allocator selects an input port and the flit is sent through the output port, the flow control unit issues a credit signal to the upstream router together with the used VC to inform the preceding router that a flit was removed from the input buffer. If the flow control unit receives such a credit signal, it increases the credit counter of the corresponding output port and VC again. If all credits are depleted for a particular port and virtual channel, no further flits are sent this port and VC until credits are available.

Routing

The NoC uses a table-based source routing scheme, offering a flexible but deterministic and simple re-routing capability. In the baseline version, the routing supports eight ports at each router to support different topologies, like mesh, ring, star, tree, and any other topology that can be created of routers with up to eight ports. In a 2D mesh topology, four ports can be used to connect peer routers, while the remaining ports can be used to connect nodes (e.g. processors, memories, off-chip interfaces). It is possible to use more or less ports. When using less ports, the encoding of the route might require fewer bits in the header and thus allow to decrease the link width.

The re-routing capability can be used for load balancing, adaptation to errors (e.g. take alternative path), or to avoid interference (e.g. dynamic re-routing by the control layer). For this, the routing table can be modified by the control layer. Source routing is selected, as it allows a simple per source re-configuration of the used paths. For distributed routing schemes, the routers have a table or algorithm to derive the output port based on the destination address. To re-route the path from a sender to a destination, one would have to change the table in all routers on the new and old path. This would increase the time needed to re-route a path and, hence, also increase the load and power consumption, as multiple re-configuration packets are needed (e.g. one for each router on the new and old path). Additionally, there are three variants for deriving the output port when using a distributed routing scheme: (1) only using the destination address, (2) using destination address and incoming port, or (3) using destination and source address. For the first two options, only a small table is needed inside the routers. However, route reconfiguration can only be done for all streams sharing the output or input and output. For the third options, the size of the router is increased, but, similar to source routing, the re-configuration can be done per sender. Hence, source routing is the most promising solution.

In the baseline version, the route is stored in 30 bit using a run-length encoding as shown in Figure 5.4 (cf. Section 2.1.2). Each run is composed of a port number (p_i) and a count (c_i). Together, they represent a sequence of hops during which the packet takes the same output port. The individual fields of the route are:

- hcnt: number of hops remaining in current run (3 bit)
- pi: port for run i (3 bit)
- ci: number of hops -1 of run i (3 bit)
- Ud: destination port (3 bit)

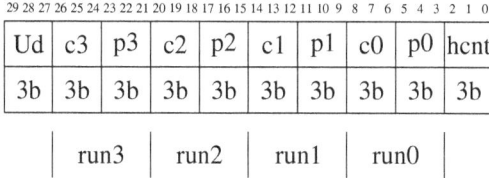

29 28 27	26 25 24	23 22 21	20 19 18	17 16 15	14 13 12	11 10 9	8 7 6	5 4 3	2 1 0
Ud	c3	p3	c2	p2	c1	p1	c0	p0	hcnt
3b	3b	3b	3b	3b	3b	3b	3b	3b	3b

| run3 | run2 | run1 | run0 |

Figure 5.4: *Route encoding of the data layer.*

The route field is dynamic and changed by the switches as the packet is forwarded. In the switches, the route is processed as follows: At each hop, the next port is found at $p0$ and $hcnt$ is decremented by one. If $hcnt$ becomes 0, there will be a turn in the next switch. When a turn occurs, the route fields ($p0c0$ through Ud) will be rotated by six bits to the left, where $hcnt$ is replaced by $c1$ before the rotation. This way, the next switch will see the next port in the $p0$ position. Additionally, as the current output port is at $p0$ and the switch calculates the output for the next switch, the route computation step can be in parallel to the switch allocation of the router pipeline (cf. Section 2.1.6).

This encoding allows addressing different nodes in a 16×16 NoC with up to six possible paths to a destination. As a single hop count can only achieve up to eight hops, up to two runs are needed to cross a single NoC dimension. For a NoC with a smaller diameter (e.g. 8×8) more possibilities exist or the size of the hop count fields (and thus of the route) can be decreased. Similar, for a 2D mesh with up to five ports per router, the directions and destination ports can be encoded in two bits (using a 2-bit clockwise router port address encoding [154; 155]), thus reducing the route field to 25 bit.

To identify the sender of a message, there is an additional field in the first flit of a packet denoting the source port (Us). When a packet is processed at the destination and, for example, a reply is to be sent back, the sender can be identified by the following approach:

1. swap Ud and Us
2. mirror the port in each run (e.g. E$\leftrightarrow$W, N$\leftrightarrow$S)
3. reverse the route field
4. set $hcnt$ to $c0$ (after rotation)

For this to work with shorter routes (less than four runs), the original sender has to put Us into the corresponding port field (p1–p3). Hence, the source of a packet can be derived from the route and no additional source address field is needed (cf. Section 2.1.2).

Error checking

The data layer supports only basic error checking mechanisms to highlight the interaction with the control layer. However, further error checking can be implemented on the software level (cf. Section 1.4) or through additional hardware extensions [34; 80; 181; 182; 184; 185; 186; 240].

We use header extensions and router extensions for error detection and handling. In each flit header there is a parity bit protecting the flit header. The parity bit protects the header information needed for the routing. This allows a quick and simple verification in the routers to handle corrupted flits early. That way it prevents erroneous flits from disturbing the router functioning (e.g. prevent a miss-routed tail flit, closing the wrong channel and leaving the corresponding one unusable) [181; 186]. Additionally, there is support for a CRC checksum for each flit, which can be enabled on a per flit basis via a flag in the header. The current implementation offers a CRC-8, allowing a hamming distance of three for up to 247 bit assuming low, constant random independent bit error rate (BER) [116]. However, as the current network packet format has some spare bit available, higher order CRCs are possible. The CRC covers the whole flit, including payload and header. The CRC is only used at the destination NI to support a simple automatic repeat request (ARQ) scheme [197]. That way, the routers do not have to re-calculate the CRC when the route is shifted.

In addition to the data and header protection, there is a 4 bit counter in each flit. This is increased by the sending network interface for each packet (i.e. all flits of a packet have the same value). When the counter reaches the maximum value, the NI restarts it with zero for the next request. The counter is used (at the receiving NI) to detect packet loss, data duplication, and if a tail belongs to a header flit (cf. Section 1.4).

In addition to the additional header information, the routers have some basic support for health checks and an interface, which can be accessed by the control layer. Through the interface, the data layer routers can report errors (e.g. detected through the parity bit). So far only rudimentary health checks are supported, while more sophisticated solutions are possible. The control layer uses this information to decide if a link might be corrupted and re-routing is needed. As a health check, the switch uses simple progress monitors and sanity checks for the input queues. That is, if there is only a single input queue requesting for an output (and there is credit available), but the switch allocator is not selecting this port (e.g. no port at all or a port without any request), an error is reported. Similarly, if a buffer with flits pending is not generating any requests for output ports, there is an error. In addition, if there are requests pending but there is no progress due to missing credits, a (permanently) blocked channel can be identified.

Besides the monitoring and reporting, the interface can be used to flush the queues and the state of routers (e.g. VC-, port-, or router-level reset) and, if applicable, re-configure parameters of the router. This allows to fully control the behaviour of the network routers through the control layer.

Network Packet Format

Messages in the NoC are transferred as packets. The logical entity of packets are flits with a size of 160 bit whereas the physical transmitting occurs over phits. In the baseline version a flit consists of four phits with a phit size of 40 bit, which are sent in a chain through the network. We differentiate four types of flits: *single*, *head*, *body*, and *tail*, with the encoding as shown in Table 5.1. This leads to the packet format shown in Figure 5.5. It allows three types of packets: a single flit packet (only a single flit), a two flit packet (head and tail flit), and a packet with an arbitrary length (head flit, multiple body flits, and a tail flit). However, in the baseline version the NoC only supports packets with a single or with four flits. For arbitrary length packets an additional length field or flit level counter would be needed in the packet or flit header to detect the loss of a body flit.

Table 5.1: Encoding of the flit types (FT) of the data layer.

Flit Type (FT)	Code
Single-FLIT	"10"
HEAD_FLIT	"11"
BODY_FLIT	"00"
TAIL_FLIT	"01"

Based on this and the descriptions above, Figure 5.6 shows the format of the head flit and thus of a single flit packet. And 5.7 shows the format of the body and tail flits. In the figures, the flits are split into four phits. The phits can be used to hide the flit arbitration latency. For this, the link width is adapted to the size of a phit and the phits of a flit are sent in a pipelined

Head Flit	n Body flits ($n \geq 0$)	Tail flit (opt. for $n = 0$)
160 bit	$n \cdot$ 160 bit	160 bit

Figure 5.5: Data layer packet format.

manner. For efficient forwarding, it is required that the first phit of the head flit contains the information necessary to select the destination output port for the next hop (i.e. that the complete route must gits into the first phit). However, other serialisation schemes are possible too, as for example using smaller link widths and transmitting a phit in multiple cycles or transmitting the whole flit at once.

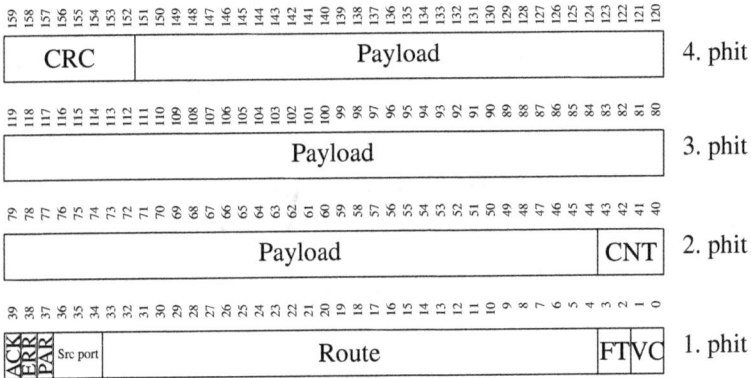

159 158 157 156 155 154 153 152 151 150 149 148 147 146 145 144 143 142 141 140 139 138 137 136 135 134 133 132 131 130 129 128 127 126 125 124 123 122 121 120

CRC	Payload	4. phit

119 118 117 116 115 114 113 112 111 110 109 108 107 106 105 104 103 102 101 100 99 98 97 96 95 94 93 92 91 90 89 88 87 86 85 84 83 82 81 80

Payload	3. phit

79 78 77 76 75 74 73 72 71 70 69 68 67 66 65 64 63 62 61 60 59 58 57 56 55 54 53 52 51 50 49 48 47 46 45 44 43 42 41 40

Payload	CNT	2. phit

39 38 37 36 35 34 33 32 31 30 29 28 27 26 25 24 23 22 21 20 19 18 17 16 15 14 13 12 11 10 9 8 7 6 5 4 3 2 1 0

ACK ERR PAR	Src port	Route	FT VC	1. phit

Figure 5.6: Single and head flit format.

The goal was to support a packet with 64 B payload. This is a typical size to transmit cache lines for a multi processor interconnect. For a packet with four flits the sum of all payload bits is 538 bit or 67 B. Hence, three additional bytes are available to attach data related information like cache line status or local process id. Additionally, if no error checking is needed, the CRC check can be disabled (by setting the *ERR*-field in the head flit to zero), allowing 570 bit or 71 B of payload data. For a single flit packet the packet format allows 109 bit with error checking or 117 bit without error checking of payload data.

159 158 157 156 155 154 153 152 151 150 149 148 147 146 145 144 143 142 141 140 139 138 137 136 135 134 133 132 131 130 129 128 127 126 125 124 123 122 121 120

CRC	Payload	4. phit

119 118 117 116 115 114 113 112 111 110 109 108 107 106 105 104 103 102 101 100 99 98 97 96 95 94 93 92 91 90 89 88 87 86 85 84 83 82 81 80

Payload	3. phit

79 78 77 76 75 74 73 72 71 70 69 68 67 66 65 64 63 62 61 60 59 58 57 56 55 54 53 52 51 50 49 48 47 46 45 44 43 42 41 40

Payload	CNT	2. phit

39 38 37 36 35 34 33 32 31 30 29 28 27 26 25 24 23 22 21 20 19 18 17 16 15 14 13 12 11 10 9 8 7 6 5 4 3 2 1 0

Payload	PAR FT VC	1. phit

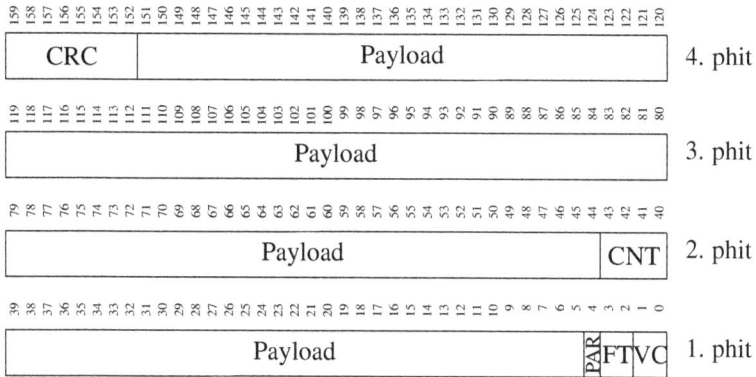

Figure 5.7: *Body and tail flit format.*

5.3.3 Control Transport Layer

The control transport layer is used for the QoS control protocol. In general, the control layer consists of three components: the clients, the RM (which might be included in the clients for a decentralized approach), and the transport of the control messages (cf. Section 4.5). The clients and the RM are implemented in the NIs or nodes (cf. Section 5.3.4) [122; 125; 223]. This section focuses on the transport of the control messages. While the control messages can be sent on a (high priority) virtual channel, we use a dedicated control layer (cf. Section 5.3.1). However, the assumptions for the dedicated layer can also be used to simplify the virtual channel approach (e.g. decreasing the buffer sizes). From the QoS perspective, the control layer has two main functions: *admission control* (e.g. allowing or denying access to the NoC) and *online reconfiguration* (e.g. adapt rate-limiters and routes based on the system state) as described in Section 4.5.

Compared to the data layer, the load on the control layer is low. Typically, a sending node will issue a single request (i.e. packet) on the control layer and then send many packets on the data layer. The online reconfiguration, based on, for example, monitoring informations or errors, also occurs slowly. The details on the protocol are presented in [122; 125; 223]. The lower performance requirements of the control layer can be used to simplify the architecture of the control layer and enable a more power efficient implementation compared to the data layer. This leads to a trade-off between low power and low latency during the design time of the control layer.

The frequency of control messages also depends on whether a centralized or decentralized approach will be used. Additionally, this decision can influence the complexity of the used interconnect. For example, for a centralized approach, all requests go to the same, predefined destination (the RM). Hence, the routing and arbitration stages in (request) network routers can be simplified, thus saving area and pipeline stages in the router logic.

In the following, we will develop an exemplary, NoC-based architecture for the control transport layer also supporting the distributed control layer. It bases on the data layer shown in Section 5.3.2, but using some simplifications, as, for example, using no virtual channels.

Message and Packet Format

To optimize the architecture for the control messages, we first define the content and requirements of these messages. As the control messages depend on the selected protocol variant, we make some exemplary assumptions on these. We can distinguish five different message kinds at the control layer: *request, acknowledge, information/status, configuration*, and *handshaking* [122; 125; 223]. The *request* and *acknowledge* messages are short messages, mainly for admission control. A sender asks the RM for access to the network and the RM responds with a *yes* or *no*. For this, the RM needs to know which sender is issuing the request and what the sender wants to do. The latter is only important, if a sender can take different actions. If there is just a single scenario for a sender, e.g., a sender only uses the control layer to obtain access to a single shared resource, the latter can be omitted. For *information/status* and *configuration* messages we need more data, as they transport measurements or configurations. Next to these NoC-RM protocol dependent messages, the *handshaking* messages can be used for end-to-end message acknowledgement and error handling of control messages. The receiver of a message checks the CRC and parity bit to verify the message. Based on the result an acknowledge or error is sent back to the sender enabling an end-to-end verification of the transmission following a *stop-and-go* ARQ scheme [197; 207]. To reduce the number of needed wires (e.g. to limit the sizes of a flit and phit), we use single flit messages for request/acknowledge and two flit messages for the other. Together with some general packaging overhead, we can come up with the flit format as shown in Figure 5.8.

This figure shows that the packet and flit format allows serialising a flit as multiple phits. For a router with a multiple stage pipeline, this serialisation induces no or only a low additional latency for single flit packets when the number of phits is smaller than the number of pipeline stages, as the subsequent phits can be transmitted while the head phit goes through the pipeline stages (i.e. pipelining of phit transfer and processing in routers). In general, the phit size could be selected even smaller. However, as soon as the first phit does not contain all information needed for the routing, the processing and transmission of phits cannot be fully pipelined. Hence, this will lead to an additional delay.

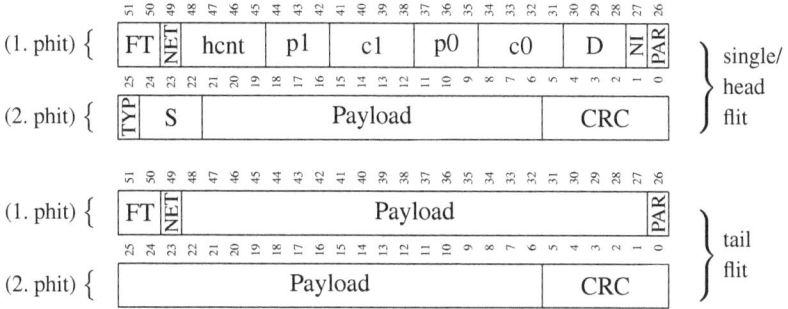

Figure 5.8: *Control layer flit format.*

The figure also shows the different data fields of a flit/phit. The *NET* and *FT* (flit type) fields denote the flit and message kind. Together they distinguish eight different message kinds as shown in Table 5.2. The meaning of certain fields and the content depend on the message kind.

FT	NET=0	NET=1
00	Request (single flit)	ACK/NACK (single flit)
01	Handshake (single flit)	Handshake (single flit)
10	Info (head flit)	Config (head flit)
11	Info (tail flit)	Config (tail flit)

Table 5.2: *Control layer flit and message types.*

The route is similar as for the data layer and stored in the $hcnt, p1, c1, p0,$ $c0,$ and D fields. In contrast to the data layer, the control layer uses only two runs and thus longer hop count fields (i.e. four instead of three bit). Hence, there are only two possible paths between a sender and destination (i.e. XY and YX routing). Additionally, to the route, we have a *NI* field, denoting whether the message is from/to the network interface (e.g. node) or the router itself. As soon as the destination is reached (e.g. the port/direction field points to a local port), the *NI* field is checked.

The *PAR* field is a parity bit for simple error checking of the header (or of the flit type for tail flits) similar as for the data layer (cf. Section 5.3.2). Next to the simple check of the header, an additional *CRC* is used for end-to-end flit-level error checking. The control layer provides 6 bit for a CRC-6 covering header and payload. The CRC-6 allows a hamming distance of three for the 52 bit of data assuming low, constant random independent bit error rate (BER) [116].

The *TYP* field denotes for a response packet, if its an *ACK* or *NACK*. For a request packet it denotes whether the request is a write or read access.

The *payload* then carries the additional information needed for the various protocol steps/messages. The actual payload varies based on the selected protocol [122; 125; 223]. In a single flit packet there is a 16 bit payload, accounting, for example, for a 12 bit bit destination address and the remainder for additional information (e.g. the requested load) or a scenario id [122; 125; 223]. For handshaking and error handling of the control messages ($FT = 01$) the payload contains the destination address of the received message (as an identifier of the message) and a bit denoting if the messages was received correctly or not. Due to the route encoding, the sender of a request can be identified by the source port field (S) and the route (cf. Section 5.3.2). However, if the header of a control message is erroneous, the sender cannot be identified correctly. In this case, the receiving interface must notify a higher level protocol or the sender must implement a timeout to recognize a lost request or response.

Router and Link Design

The control layer needs no virtual channels, uses small packets (i.e. one or two flits per packet), and has a low load. These properties allow to use a less complex router compared to the data layer.

As there are no virtual channels, the router has only a two stage pipeline with a combined route computation (RC) and switch access (SA) stage and a switch traversal (ST) stage. Additionally, there are no multiplexers between the input and the crossbar, but the input buffers are directly connected to it, simplifying the arbitration. Similar as for the data layer, the port for the current router of a flit was already computed at the previous router or the sending NI. During the RC stage, the router adapts the hop count field and, if necessary, rotates the route fields for the next hop. As the control layer has no virtual channels there is the possibility for a protocol deadlock for certain protocols [122; 125; 223]. An example for such a deadlock is a cooperative, exclusive access protocol, where a sender has to release the resource before the next sender can get access. In such a design, it must be guaranteed that release messages will always arrive at the resource manager (or other clients). Hence, the requests and releases should not experience high backlog at the resource manager (or clients), such that the blocking propagates to the network. For this, the backlog of requests can be formally derived and the buffer sizes (at the RM or clients) set accordingly (e.g. using the analysis from Section 3.4). To prevent such unfavourable designed systems, the control might supply additional virtual channels or bypass channels/buffers for certain message kinds.

To save area and power, the router uses single flit buffers. While this can potentially lead to the case where a packet is split across multiple routers when contention occurs, the low load of the control layer makes this scenario rare and diminishes the problem of blocking propagation. The link width equals the phit width, where each flit has two phits as already shown in Figure 5.8.

Next to the mentioned basic router parts, the control router has an additional interface to the routers of the data layer. This interface helps to monitor and control the data layer routers (cf. Section 5.3.2). Through this interface, the control layer can access and modify all configuration parameters of a router and also remove flits from the queues (e.g. flushing blocked ports). Additional, the NoC-RM or higher level software can collect monitoring information on the occurred errors, load of a port, and potentially blocked queues through this interface. This information can then be used by the NoC-RM to adapt routes (e.g. for load balancing or fault tolerance).

5.3.4 Network Interface

The network interface (NI) connects the computing blocks or nodes to the network routers. It is responsible for packaging of any requests into a NoC compatible format and sending it over the NoC. It also isolates the nodes from the network and provides admission and rate control and, thus, provides the client interfaces for the control layer. Figure 5.10 shows a high level block diagram of the NI of the proposed architecture. In the figure, red blocks denote parts that are added or extended to a typical NI to support the control layer. In the NI we can differentiate between up to three layers. The first layer translates the local node protocol to the internal NI protocol. That way the NI internals are independent of the actual implementation of the node. The second layer comprises the functions of the NI, which includes the

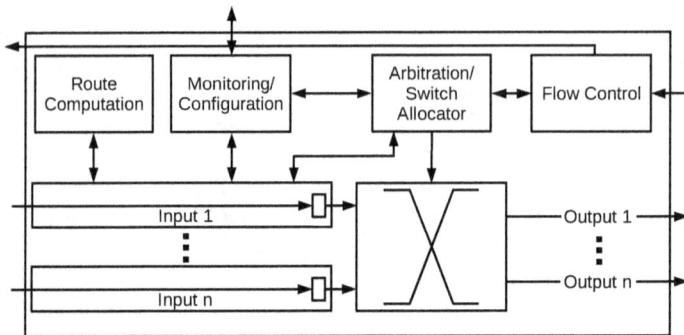

Figure 5.9: *Router architecture of the control layer.*

packaging of messages in packets and flits, as well as control and monitoring functions. And the third layer is the output stage to the network on chip. For a simple NI it only contains buffers for the communication with the NoC. In more complex designs, it might also contain components for clock domain crossing and an additional protocol translation if the NoC and NI format differ (e.g. to adapt different link widths).

Between the different architectures of the control layer the NI mainly differs in the number of physical output channels, i.e., if the control layer is connected to an own physical channel or a virtual channel of the data layer. In the following we provide details on the individual components of the NI.

Packetization

The packetization and routing converts the messages of a node to packets and flits. For this, there is an address translation and routing based on an address translation table. Figure 5.11 shows an example for such a table. A local (physical) address is converted into a remote address (e.g. on another processing core) using a pre-defined route through the network. The route includes the path and virtual channel. Additionally, the table has control fields denoting if the node has write or read access to the accessed address. This basic table is extended by entries for the control layer. For each entry, there are fields denoting if the requests needs to be synchronized by the control layer (*Sync?*) and if it was already granted (*Ack?*). This can be further extended by a throughput field (BW) denoting the required throughput. To achieve a low overhead for the control messages, an *ID* field is added for the control layer. This ID can be used by the control unit to distinguish

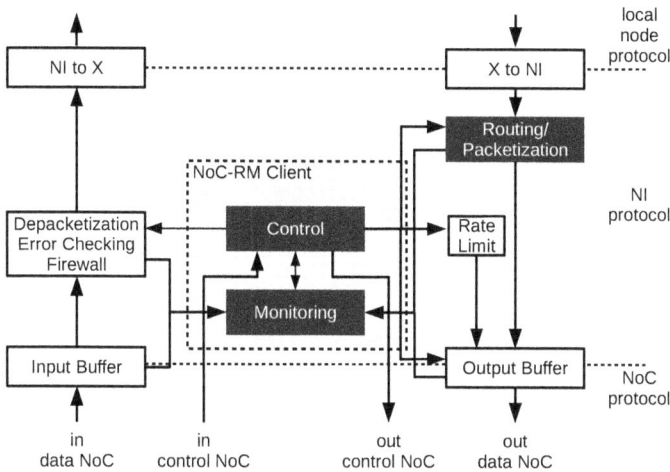

Figure 5.10: *Network interface with NoC-RM extensions.*

certain scenarios the sending node belongs to. It can be used, for example, to only transmit the ID to the resource manager (or other clients) instead of the destination address and other requirements. Basically, it compresses the local address, route, remote address and optionally the throughput requirement to identify a request. Additionally, the settings for the rate control can be stored in the table. That way, the NI can locally reprogram the rate control for already acknowledged destinations. These fields can be merged with the control unit or global state. In such a case, only an identifier needs to be stored to index the other tables. In simple scenarios this identifier might just be the route, remote address, and potentially the local address. Hence, no extensions to the basic address translation would be needed.

The table can be modified by the control unit. This can be done based on received control messages (e.g. from a supervisor node, resource manager, or software running on the node) or on monitoring events. These modifications can be done during the run-time, allowing the system to adapt to changes (e.g. new applications, faulty network paths). During design time, the designer decides which components are allowed to issue such (re-) configuration messages to the control module and if this list can be changed during run-time and by whom.

If there is an access to an address, which is not present in the table, the access is reported and the control unit can directly block or drop the access

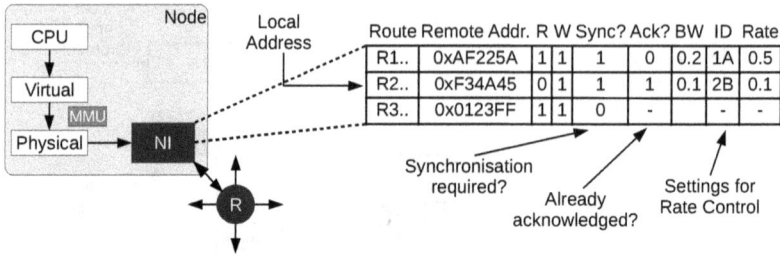

Figure 5.11: Address translation table of the NI.

or send a message to a supervisor node or the resource manager to request the entry or desired action.

Depacketization

The depacketization translates the (incoming) network packets to the NI protocol and forwards the data to the corresponding module. This module also contains simple error checking and firewall mechanisms.

The error checking is done on flit level, e.g., by parity bits and a CRC check (cf. Section 5.3.2). The parity bit is a simple check for the packet header and is also used by the router for a quick detection of 1-bit errors. The CRC check covers the whole flits, including header and payload. That way it can detect misrouted or corrupted flits. It also includes a check for the (sequence) counter in the flits, verifying if the counter was incremented correctly. That way, the NI can check, for example, for the correct sequence of data or flit loss.

Additionally, the depacketization module contains a table containing a list of possible senders and their access rights (similar to the address translation table). If an access is not allowed by the table (e.g. wrong access type or unknown sender), an error is reported to the control unit (via the monitoring). Based on the decision of the control unit, the erroneous packet can be dropped (silently), rejected (non-silently), or forwarded to the local node. The content of this table can be modified through the control unit, similar as for the address translation table.

Rate Limiter

The rate limiter can enforce a maximum rate with which the local node can emit traffic to the network. In the baseline design it is a simple token bucket shaper [153; 216]. However, other design are also possible as, for example, dynamic counters or *l*-repetitive functions [133; 160; 161; 217]. Note, if more complex rate limiting mechanisms are used, the tables containing the allowed rates and the control messages might need to be extended. The rate limiter can be modified by the control unit during run-time. That way, the system can adapt to changed conditions and enforce the assumed rates for the scenario selected by the control layer.

Noc-RM Client

The NoC-RM client comprises the different modules needed for the control layer. It consists of two modules: a *monitoring* module to collect local information and a *control* module implementing the NoC-RM protocol and controlling the other modules in the network interface.

The monitoring module enables the control unit to monitor the actions of the other modules. For the NoC-RM, its main function is to monitor the packetization module and recognize accesses of the node to the NoC. When the node accesses the address translation table with a local address and the requests requires a synchronization ($Sync? = 1$) and is not acknowledged ($Ack? = 0$), the request is intercepted and forwarded to the control module. This information is then used by the control module to detect outgoing transmissions and, if necessary, conduct the global resource management.

Next to this basic functioning, the monitoring module was extended to provide further information to the control module and the NoC-RM. For this, the configuration messages of the control layer can be used to configure a trigger value in the monitoring module or to request the current state of a certain value. While the latter corresponds to a simple read of the state, the former can be used to automatically generate information messages when a threshold is reached. In the current design, three monitors are supported. First, the monitor can be used to check the fill level of the output buffer. With this information, the NoC-RM protocol can check if the NoC accepts sufficient throughput. For this, the monitor can generate a notification when buffer overflow occurs to inform the higher level protocol (e.g. the resource manager). Similarly, the monitor can be enabled to send a notification when the depacketization and error checking module observes an erroneous flits. This way, the NoC-RM can use the end-to-end flit level CRC to detect corrupted links. And finally, the monitor can check for buffer overflow of the input buffer. If the input buffer overflows a notification can be generated. This information can be used to detect a malfunctioning node that no longer accepts incoming messages. In such a case, the NoC-RM protocol might decide to exclude the malfunctioning node from the network by denying futures accesses to it.

The control module is responsible for basic control functions (e.g. configuring the address translation tables) and the logic for the control layer. This includes the negotiation of access through the control layer, re-configuring of the local parameters (e.g. rate-limiting), and notifying the resource manager (or other senders) when a transmission has finished. Based on accesses to the network by the node, the control unit emits request messages via the control layer. For example, it asks the resource manager or other clients, if a new connection is allowed, what setting for the rate limiter should be used, and which route should be selected [122; 125; 223].

The working of the control module depends on the selected architecture (i.e. centralized or decentralized approach) and the protocol. When a sender tries to access the NoC (i.e. access to the packetization module / address translation table), the NoC-RM client will detect and intercept the request. Based on the accessed destination and an optional identifier the client identifies the request (i.e. transmission). For each transmission requiring synchronization the current state of the needed resources is stored in a table at the client (or in the address translation table). For a centralized approach this is only be a *allowed* or *not allowed* (i.e. the *ACK* field in Figure 5.11) together with a value for the rate limiter. For the decentralized approach, the tables stores the state of each required resources (e.g. current reserved load of each router on the path). Hence, the amount of information stored strongly depends on the version of the implemented protocol. The size of the table depends directly on the number of different connections and their parameters (connection settings and system modes) as well as the number of senders that require synchronization. Hence, there is a trade-off between a fine-grained control for improving performance and the necessary resources for the control layer (e.g. size of the state tables).

If a negotiation (with a RM or other clients) is necessary, the control module generates the corresponding protocol messages and delays the intercepted network access until the negotiation is finished. Additionally, due to changes of the system state, re-configuration or status update messages can arrive at the control module. These are used to update the local state table and possibly to change the settings of the rate limiter, monitoring module, or delaying of further accesses.

Figures 5.12 to 5.14 show block diagrams for the architecture of the resource manager module, the client for the centralized approach, and the client for the decentralized approach.

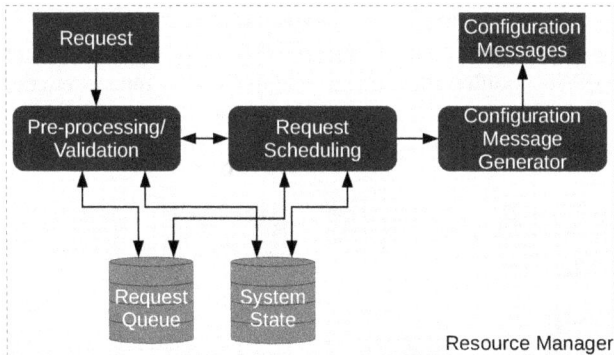

Figure 5.12: *Resource Manager (centralized approach).*

Figure 5.12 shows the resource manager (RM) module for the centralized approach. If the RM is implemented as a dedicated node, this figure amplifies the control modules of the network interface (for the RM). The RM mainly consists of three modules and some memory to store requests and the system state. The pre-processing and validation module checks if a request is valid. Additionally, it can be used to filter out requests of a sender. That is, if the control layer protocol supports to gradually increase the requested load of a network node, it can remove unnecessary requests from the request queue. For example, if a node first requests 10 % of link load followed by a request for 20 % load (before the first one is selected), the RM can safely switch to the 20 % case without reconfiguring the system for the 10 % case. This filtering of messages strongly depends on the used protocol and might not be allowed for all cases. The decision on this is made by the system designer during the design of the protocol. Valid requests are then stored in a request queue. This also includes release messages of resources. From the request queue, the *request scheduling* selects the next mode for the system accounting for the current global state. For this, different scheduling mechanisms and other constraints can be implemented as discussed in [122; 125; 230]. If the next mode is selected, it is forwarded to the *configuration message generator*. This module generates the needed configuration messages for the whole system and emits them. All modules can be implemented in software, hardware, or a co-design. However, a reasonable choice seems to be to implement the *configuration message generator* as a hardware accelerator and the other modules as software on a basic processor or specialized processing

core. That way, the *configuration message generator* can generate the configuration messages without high timing overhead (e.g. one message per cycle),
while the other modules allow an easy adaptation by the system designer or
programmer. Additionally, this design allows the *request scheduling* to derive
the next state while the configuration messages are generated in parallel.

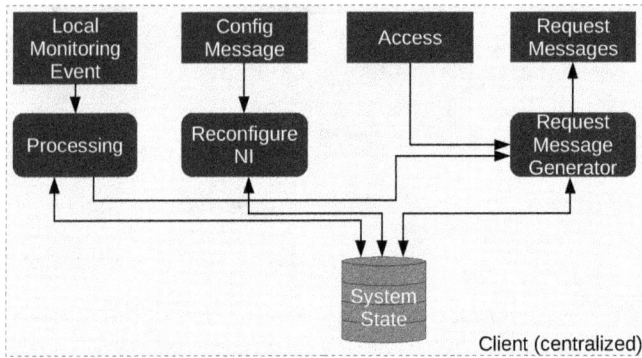

Figure 5.13: *Client (centralized approach).*

Figure 5.13 shows the client module for the centralized approach. The
figure amplifies the control modules of the network interface for a client
node using the centralized control layer. As the main logic is handled by the
resource manager, the client module is simple. The client mainly consists of
three modules and some memory to store the state. There are three events
that can trigger the client module: a *local monitoring event*, an arriving
configuration message, and an *access* (including releases of the local node).
If a monitoring event arrives, it is processed together with the current state.
The processing unit decides, if the monitoring event needs to be reported to
the resource manager. If so, it generates a request message. Otherwise only
the state is updated. When the client receives a configuration messages, it
triggers a reconfiguration of the local network interface and updates the state.
If the local node is doing an access that needs synchronisation, the client
traps the access (e.g. the address translation table delays it) and generates a
request to the resource manager. After the resource manager has processed
the request, it replies with an acknowledge or configuration messages. As
soon as these arrive (as a configuration message) at the client, it reconfigures
the NI, i.e., it unblocks the access in the address translation table and, if
necessary, programs the rate limiter.

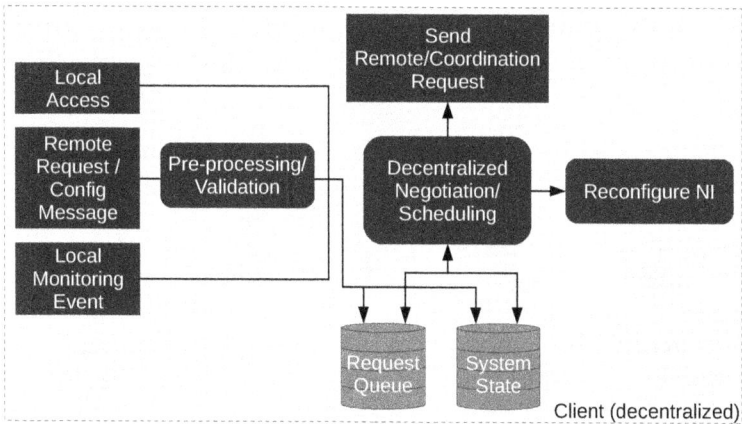

Figure 5.14: Client (decentralized approach).

Figure 5.14 shows the client module for the decentralized approach. The figure amplifies the control modules of the network interface for a client node using the decentralized control layer. As there is no centralized unit, multiple clients have to conduct a distributed scheduling and decision-making protocol to select the next valid state. The client mainly consists of three modules and some memory to store requests and the state. There are three events that can trigger the client module: a *local monitoring event*, an arriving *remotes request or configuration message*, and a *local access* (including releases) of the local node). All events are stored in the request queue and the state module. For remote requests there is a validation to check if a request is valid. Based on the request queue and the state, the client conducts a distributed scheduling and decision-making protocol with other involved clients. For this, it possibly exchanges multiple messages with other clients. As soon as all clients agree on a new state, all needed clients reconfigure their NI [122; 125; 223].

The presented control modules are examples for the general architecture of these modules. The actual protocol and working, and thus the needed messages and actions by the control layer are defined by the designer [122; 125; 223]. Hence, the actions (e.g. number of messages exchanged between the clients or between a client and the RM) can vary between different designs. However, the presented designs cover a broad range of different protocols, e.g., all protocols presented in [122; 125; 223], as the main logic is done through the (SW) protocol and not fixed hardware components.

Buffers

The input and output buffer of the NI mainly buffer NoC packets (e.g. flits) to relief the other NI modules from low level and physical layer NoC access handling. Additionally, they can be used for clock domain crossing and protocol translation, when the NI internal communication uses a different (physical or logical) protocol than the NoC. The buffers have an interface to the monitoring module and the output buffer has an additional interface to the control unit. Using this interface, the control unit can delete data from the output buffer and inject own data packets.

Protocol Adapter

For the connection to the local node, the NI has two protocol adapters (*NI to X* and *X to NI*). These translate between the node protocol (e.g. AMBA) and the NI internal protocol. That way, the NI internals can be designed independent from the actual node design. If the node, or more precisely its communication protocol, changes, only the protocol adapters need to be adapted.

5.4 Summary

In this chapter we presented a NoC architecture that provides support for the NoC-RM. It provides two networks, one for the data layer and one for the control layer. For the data layer it uses a NoC with virtual channels supporting up to four priorities. The virtual channel assignment is fixed from sender to destination, i.e., the virtual channel (VC) is not changed in the network routers. The control layer can be used to adapt the VC used by a network node or traffic stream during run-time. For this, the control layer can adapt the address translation table (e.g. routing table) inside the network interfaces. The architecture uses a dedicated interconnect for the control layer to isolate the control and data messages from each other. This way, the control messages do not contribute additional interference to the data layer, which might limit the benefits of the NoC-RM. The control interconnect uses a NoC that is optimized to the control layer protocol. Hence, the size of the routers and the link width is smaller than for the data layer. With this, the proposed architecture for the control layer solves the requirements from Section 5.2. That is, the dedicated interconnect provides a low latency for the control messages and avoids direct interference with the traffic on the data layer.

Besides the transmission of the control messages, the control layer architecture supports additional interfaces to the data layer. These are used to monitor and supervise the data layer routers. With this information, the control layer can efficiently adapt the system to changes and errors.

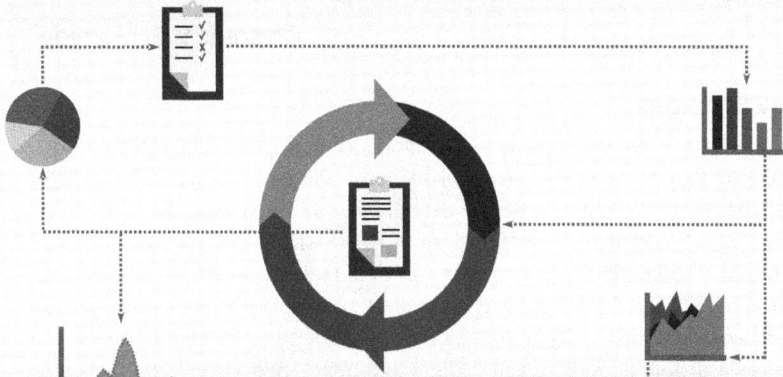

6. Evaluation

In this section we evaluate the network-on-chip (NoC) architecture supporting the NoC-RM from Chapter 5 against other architectures. For this, we first evaluate the performance of the presented architecture using different control protocols against a classic prioritization scheme and a virtual channel based control layer using a network simulation. This is followed by an overview on the estimated hardware overhead of the control layer. The chapter is partially based on the work published in [223].

6.1 Simulation Framework

The simulation based evaluation is carried out with the *OMNeT++* event-based simulation framework and the *HNOCS* library [12; 32; 226; 227]. In the experiments we compare a standard priority based NoC with a 2D mesh topology and virtual channels (VCs) against the proposed architecture for a decentralized NoC-RM approach from Section 5.3. An evaluation of a centralized architecture can be found in [122; 230]. The standard network uses wormhole switching where a data packet consists of eight flits. The network routers support a two stage priority-based arbitration and use a 4-cycle pipeline. For all architectures, we use a XY source routing scheme as a baseline for the path allocation. The buffer depth for each input VC corresponds to two flits. Additionally, we assume a 1-cycle link traversal. The interconnect frequency is set to 500 MHz. For the network traffic, we use memory transfers since memories are frequently a common hot spot module in MPSoCs [171; 172].

6.1.1 QoS Schemes

For the evaluation, we consider different distributed protocols of the control layer: an exclusive access protocol for temporal isolation (TIS), adaptive load distribution (ALD) of BE senders [121], and three extensions of ALD. These extensions also allow throttling of BE (SALDT and FALDT) as well as to adapt the path of safety critical senders (selective congestion control, SCC) [223]. The protocols are compared against each other and against prioritization done locally in routers (SP), see Section 2.2. The SP approach uses two different priorities in routers and hence two virtual channels. In general the SP scheme allows as many priorities as virtual channels exist. On the contrary, the approaches applying the control layer support more priorities than available VCs. Therefore, the control layer can decrease production costs and power consumption for system that need multiple priority levels. An evaluation of centralized protocols for the control layer was presented by Kostrzewa [122].

Exclusive Access (TIS)

Temporal isolation (TIS) is realized using a work-conserving scheduling based on priorities as proposed by Kostrzewa et al. [119]. For this, each synchronisation scenario has a unique token assigned. Tokens are control messages deciding about who may currently use a resource. To achieve that, they are constantly circulating between all clients controlling which senders are active in the particular synchronisation scenario. Hence, in a typical design, the clients form a logical token ring, but other topologies such as a star topology are possible. To forward the token, a client knows the address of the next sender in the synchronization scenario. The order as well as the mapping of the senders on the network nodes is prepared and calculated offline and given by the synchronisation scenario. It is static at runtime to preserve safety and predictability of the mechanism. During the platform boot-up, one specially delegated client per token is starting the token exchange. Only the sender running on the node with the client who currently possesses the token may actively use the supervised resource. If the sender will issue an access to the resource and the token is not present, the access will be blocked and stored by the client. As soon as the token will arrive the access will be resumed and the communication granted. If a client will receive the token and there is no request from the sender, it forwards it to the next candidate. In a situation where no sender has to transmit data, a monitoring mechanism should prevent the clients from forwarding the token too frequently or the design must ensure that this situation cannot occur.

Adaptive Load Distribution for BE (ALD)

Adaptive load distribution (ALD), i.e., the detouring of BE, was introduced by Kostrzewa et al. [121]. In this scheme, we apply a static path allocation to SC traffic, whereas BE traffic can choose from multiple paths to improve its latency. Whenever the network is free, BE traffic is allowed to use the optimal (e.g. shortest) path to its destination, even if it overlaps with links used by critical (SC) senders. Upon activation of a SC sender, we release the needed network resources by redirecting the BE transmissions to an alternative route. Hence, BE senders use detoured paths (e.g. possibly non-optimal routes) only when critical senders are actively using shared network resources.

To arbitrate predictably within sets of *interfering senders*, we statically assign a set of paths for each BE application with any available allocation method (e.g. [16; 72]). The set of possible paths for a BE sender does not change during runtime. In principle there are no limitations in the selection of alternative routes, besides the rule that, if a link in the path is shared with a SC sender, this sender must be capable of accepting the additional protocol overhead. However, if all paths of a BE sender are shared with SCs and all SCs are active then the BE sender will be blocked by the client similarly to the temporal isolation protocol. Additionally, a detoured path should not induce more overhead to BE transmissions than the (full) blocking during a SC transmission will. Otherwise, waiting for the SC transmission to finish (e.g. temporal isolation) is more efficient, than using the detoured path.

Each sender is equipped with a client interface supervising the accesses to the network (cf. Section 5.3.4, Figure 5.14). The client stores a table containing the current state, similar to the example in Table 6.1. It contains for each set of interfering SC senders a list with senders that are currently active (left) describing the state of the system. Upon beginning and termination of a transmission, the client of a SC sender sends control messages to all clients of interfering BE senders: *actMsg* (SC activation), *relMsg* (SC release). These messages propagate the NoC state defined by the currently active SC applications and determine the most suitable paths for the BE senders. After receiving these messages, the corresponding paths (used by SC) are blocked or released for BE transmissions. After each mode change, the BE sender will use the best (e.g. shortest) path available, i.e., a path without any active SC sender. In the example from Table 6.1 this is realized with a bit vector (array) where the position of the bit identifies a specific sender and its value the sender's state (active or disabled). This state then defines the routes that the BE sender must take (right) or if it is not allowed to send in the current state (e.g. *none* for the route). This table is stored in the system

Table 6.1: *Scenarios stored at each BE sender for ALD.*

Dest / ID	Active SCs	Used Route
0xAC1234	000	Route_1
0xAC1234	001	Route_2
...	...	...
0xAC1234	111	*none*

state table of the network interface (NI) and used to update the route of the *address translation table* (cf. Section 5.3.4). On the reception of a *actMsg* or *relMsg* message, the client updates the state. Figure 6.1 presents a simplified workflow of this protocol (used also for the mechanisms in the following sections). Actions are divided w.r.t. the criticality of the sender (SC and BE) along with messages causing mode transitions.

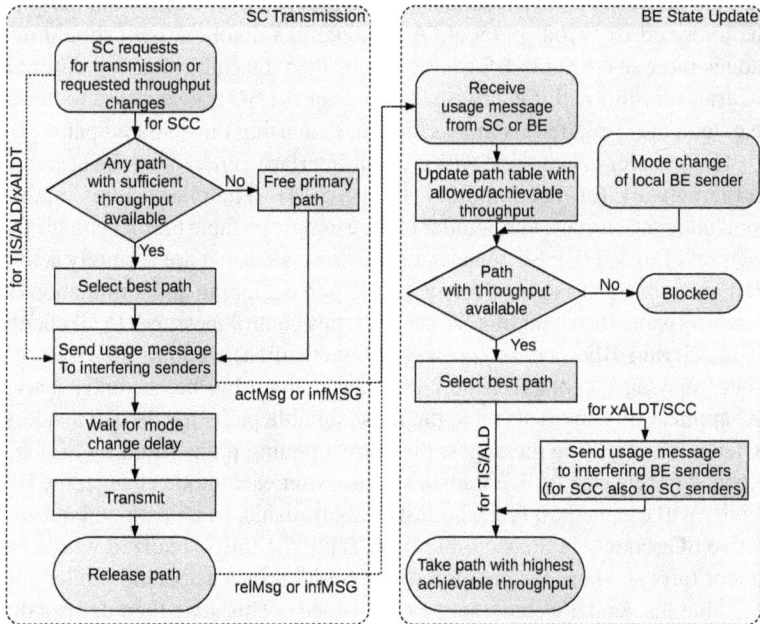

Figure 6.1: *Simplified state update protocol.*

Finally, to preserve predictability (i.e. provide sufficient isolation) of safety-critical (SC) senders, we must account for the delay resulting from time necessary to deliver control messages to BE senders and for the packets from BE transmissions to leave the interconnect on the selected path (i.e. the *mode change delay*). That is, for example, if a SC sender requires exclusive access, its client will send a message to all interfering BE nodes. The clients of the BE nodes will then adapt the routing table of the network interface and re-route the BE traffic or delay further network accesses. However, there can still be networks packets of such a sender inside the network using the old route. Hence, to avoid head of line blocking between BE and SC traffic or an overload, the SC sender has to wait until the BE packets have left the shared path. For the error free case this mode change delay $t_{modeChange}$ can be upper bounded by:

$$
\begin{aligned}
t_{modeChange} = & Inj^{ctrl}\left(|BE(h)|\right) + \max_{\forall j \in BE(h)} \left(\sum_{k \in Tasks(P^c(i \to j))} R_k^{+,ctrl} \right) \\
& + \max \left(\max_{\forall j \in BE(h)} \left(\sum_{k \in Tasks(P^c(j \to i))} R_k^{+,ctrl} \right), \right. \\
& \left. t_{reconfigure} + \max_{\forall j \in BE(h)} \left(\sum_{k \in Tasks(P(j))} R_k^{+,pkt} \right) \right),
\end{aligned}
\tag{6.1}
$$

where $BE(h)$ is the set of BE senders sharing same path with the SC sender; $|BE(h)|$ denotes the number of BE senders sharing same path with the SC sender; $t_{reconfigure}$ denotes the time of the client to reconfigure the network interface; $C_{j,ctrl}^{+}$ denotes the maximum latency of the control message to the BE sender j; $C_{i,pkt}^{+}$ denotes the maximum latency of a single packet belonging to the BE sender j; $Inj^{ctrl}(m)$ is the time to inject m control messages including de/packetization overhead; $Tasks(p)$ denotes the set of all tasks of path p (i.e. one per hop); $P(j)$ denotes the path of a stream j; and $P^c(i \to j)$ denotes the path of control messages from node i to node j. In the equation the first term accounts for the time to inject the control messages. This is followed by the maximum worst-case response time of the control message to any BE sender using the shared path (derived by the analysis of Section 3.4). After receiving a control message, the client of the BE node sends an acknowledge to the SC node. In parallel, the client of the BE node reconfigures the network interface ($t_{reconfigure}$). After the reconfiguration, we need to account for the maximum worst-case latency of the BE senders (derived by the analysis of

Section 3.4), such that all traffic of the BE nodes must have left the network. As the acknowledge happens in parallel to the reconfiguration and depletion of the BE packets, we only need to account for maximum of both.

Such overhead is often acceptable as SC senders have available time budget between the worst-case transmission latency and the deadline of a transmission (cf. Chapter 4). Safety critical senders usually have no advantage in finishing earlier than the deadline [119; 201; 217]. This allows a trade-off analysis for providing estimation of the overhead resulting from the global arbitration [223].

Adding Throttling of BE Senders

While the simple detouring of BE traffic can increase the performance for BE senders (cf. [121]), it does not account for the actual workload in the network. In many applications, the amount data, which needs to be transmitted, may vary during runtime and the state of the system, e.g., video processing as used in automated driving systems. Hence, the inter-arrival time between packets and the resulting load on the shared links can vary, such that it is not always necessary to fully avoid the interference between SC and BE. This enables to divide the behaviour of SC senders in different modes [44; 103; 157; 158; 159; 238]. For each mode, we can derive different upper bounds on the amount of traffic a BE sender is allowed to induce to shared links, without jeopardizing the requirements of SC senders. This value can be also equal to zero, effectively detouring or blocking BE senders, leading to the same behaviour as for ALD. Additionally, as the SC sender has slack and allows some interference (e.g. by the introduced protocol or BE traffic), it might share a path with BE traffic, as long as the BE senders are not inducing too much traffic (high interference). Hence, instead of detouring BE traffic every time a SC sender uses a path, a more suitable solution is to detour BE traffic only, if the interference on the shared links is too high. For doing this, we do not remove the shared path from the set of allowed paths for a BE sender as soon as an SC sender accesses it, but instead provide the clients of the BE senders with information about the maximum allowed (BE) load on the path. Then, the client of a BE sender can decide, if a detoured or throttled path will lead to a better performance and select the appropriate path.

For this mechanism, we differentiate two approaches on finding the current optimal path for a BE sender: a *semi-adaptive* approach, reducing the overhead of the needed synchronization, and a *fully adaptive* approach, optimizing the performance while at the same time increasing the synchronization overhead.

Table 6.2: Scenarios stored at each BE sender for SALDT.

Dest / ID	State of SCs	Used Route	Rate Limit
0xAC1234	000	Route_1	100%
0xAC1234	00A	Route_1	60%
0xAC1234	00B	Route_2	100%
0xAC1234	0A0	Route_2	100%
...	...	...	...
0xAC1234	AA0	Route_2	40%
0xAC1234	AAA	*none*	0%

For the *semi-adaptive* approach (SALDT) we extend the synchronization scenario from ALD with a performance indicator, based on the allowed load. Table 6.2 shows an example for this. Based on the different modes (i.e. requirements) of the SC senders in a synchronization scenario, we define the optimal assignment for BE senders. That is, instead of detouring all BE senders, we define which BE sender will use the throttled path (shared with SC traffic) and which must use a detoured path. This is done offline, based on the known worst-case requirements of SC senders in the different modes and the mapping. However, if BE and SC senders concurrently share the same link, it must be guaranteed that BE traffic never exceeds the allowed load. For this, the table contains the maximum allowed load for the sender in each mode. The traffic shapers will then be adapted by the *client* to the currently maximum allowed value (e.g. the *rate limit*). Alternatively, the worst-case behaviour of BE senders must be known beforehand (assessed offline) and known to be feasible. In that case, no rate limiters are needed. With the known assignments, we then extend the control message of SC senders to either detour or throttle a BE sender. That is, instead of sending a simple activation notification, SC now informs BE senders on the actual required throughput. Consequently, a BE sender must store the allowed route and maximum allowed throughput of this route for each state of the synchronization scenario. The state is now not only defined by a set of active senders but also in which mode the senders are (denoted by *A, B, ...* in the table). The BE sender (i.e. its client) then can check if there is a path with sufficient available throughput and if not, whether it is more beneficial to take a rate limited path or to be blocked. However, the table shows that this additional information increases the size of the data stored in the NI.

Table 6.3: Scenarios stored at each BE sender for FALDT.

Dest / ID	Route	Available Throughput
0xAC1234	Route_1	90%
0xAC1234	Route_2	35%
0xAC1234	Route_3	70%
...	...	...

The *fully adaptive* approach (FALDT) enables to also account for the actual behaviour of the BE senders in the system at runtime. For this, we extend the synchronization protocol by adding two messages between BE senders: a release message *relMsg*, which is the same as for SC senders but now denotes that a BE sender is not using a path any more, and an information message *infMsg* announcing the currently used path and the requested load on this path (cf. Table 6.4). With this additional information, a client knows the maximum BE load the SC senders allow in a synchronization scenario (from *actMsg*) and how much load of this is used by other BE senders (from *infMsg*). This allows the client to choose the path with the current optimal performance for BE senders. Hence, BE traffic will only be detoured or slowed down when necessary, improving its overall performance. Additionally, as now only the state of the allowed paths has to be stored (instead of each possible combination of active senders and their modes), the size of the stored information can be reduced (cf. Table 6.3).

Detouring of SC

So far a safety-critical (SC) sender was constrained to use a single static path. However, in a system there can exist multiple possible paths for safety-critical senders. The approaches, introduced before, solely use the available time budget of SC senders to conduct the synchronization, i.e., throttling or detouring of BE senders, between SC and BE senders on shared paths. Another option is, to also assign alternative, possibly not optimal, paths to SC senders. In such a scenario, a SC sender will use the shortest path when it is free (similar to the case of BE traffic in the ALD approach). This is done, to reduce the time SC traffic spends inside the network and to reduce the possibility for interference with other senders. If a BE or SC sender tries to access the shared path, the client of the SC sender will check if it can use a detoured path on which still all requirements of the SC sender can be satisfied. If such a path exists, the client of the SC sender will configure the SC sender to use the detoured path to allow BE traffic to use its shortest path and improve its performance. Otherwise, it will perform a detouring or throttling of BE senders as before. Hence, the client of a SC sender now also stores the system state. However, instead of being blocked, when no path with sufficient throughput is available, it will free a path, i.e., detour or block BE traffic (cf. Figure 6.1).

For this mechanism we further extend the synchronization protocol, leading to a *selective congestion control* (SCC) combining the previous mechanisms and detouring of SC traffic. The clients of the BE senders inform the clients of the SC senders in a synchronization scenario about their current state (i.e. network usage) via the *infMsg*. Hence, based on the received messages the client of a SC sender knows about the current state of the shared paths. If a SC sender wants to initiate a transmission, the client first checks for the optimal path. In this sense, an optimal path is a path, on which all requirements of the SC sender can be satisfied and all BE senders in the synchronization scenario achieve the best performance. For simplicity, we define the performance of BE traffic based on the maximum link utilization. Hence, the SC sender will take the path that minimizes the maximum link load (for the BE senders). The decision of the SC client can also include a detouring or throttling of one or multiple BE senders, as introduced before.

On the other hand, if a BE sender wants to start a transmission or changes its requirements (e.g. requested throughput), its client will select the appropriate path based on the current state. At the same time, the client releases a *infMsg* to the other nodes in the synchronization scenario. Upon receiving this message, a SC client might then re-evaluate the best path assignments and send new *actMsg/infMsg* messages to all involved BE clients. This leads to a re-configuration of the current state based on the pre-defined paths to optimize the performance of BE senders.

With this, we can summarize the different protocol messages for the approaches as shown in Table 6.4. The table shows for the different approaches (left) which client (BE or SC) is issuing the control messages (top) and to whom. The table shows that with an increased amount of information the approaches use to optimize the performance, the protocol overhead (e.g. the number of messages that must be exchanged) also increases. Depending on the implementation of the *control layer*, these protocol messages may also introduce temporal overhead (e.g. synchronization time and possibly additional interference in the network). To reduce the overhead of the *infMsgs* of all BE senders, BE senders might only emit control messages, when there

Table 6.4: Overview on the control layer protocol messages.

	actMsg	relMsg	infMsg
content	active	inactive	requested throughput
ALD	SC to BE	SC to BE	—
SALDT	—	SC to BE	SC to BE
FALDT	—	SC to BE BE to BE	SC to BE BE to BE
SCC	—	SC to BE BE to BE SC to SC	SC to BE SC to SC BE to BE BE to SC

is a significant change in its requirements. That is, it will not emit synchronization messages for each packet but, for example, only for mode changes or DMA transfers which lead to temporary different requirements towards the network. For a general purpose application, which is fully unknown, this might lead to the case, that this application might only send a protocol message if it is activated or not send any at all to reduce the overhead.

6.1.2 Use case

For the experiments, we modelled the safety-critical (SC) traffic after the real-time multimedia application performing video noise reduction as shown in Figure 6.2 [47]. This functionality is decomposed into three communicating tasks (*T1–T3*). We assume the input data to arrive via an Ethernet port (*ETH0*). The use case is mapped to a 4×4 NoC with the data-flow and mapping presented in Figure 6.3 and Table 6.5. In this use case, new data (i.e. frames) will arrive at the Ethernet port and must be forwarded immediately to the memory, to prevent a buffer overflow at the Ethernet port, i.e., package dropping. Each SC module is modelled with a traffic generator conducting 8 kB long DMA transfers, allowing to maximize the benefit from DDRAM3 2133N [9]. Transmissions are performed periodically and periods are calculated based solely on the required bandwidth including some jitter (15 % of the period) due to the data-dependent execution (video frames). We distribute the SC traffic between the DRAM ports due to the conflicting temporal (throughput) requirements and pipelining of the data processing.

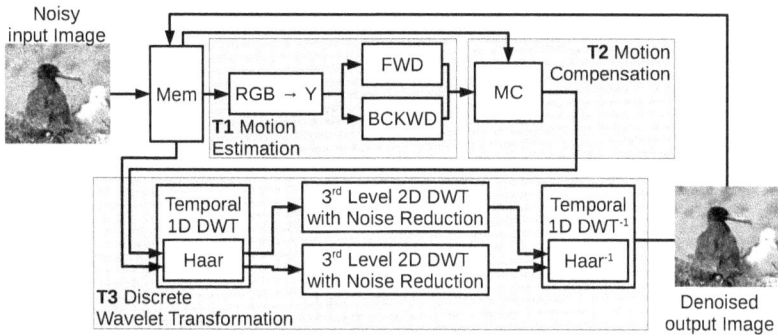

Figure 6.2: *Example real-time video noise reduction application [47].*

Additionally, to the SC traffic, several best-effort (BE) tasks are mapped on the remaining nodes in the system. These BE senders are modelled with the CHSTONE benchmark. The traces of the benchmarks are extracted using the Gem5 simulator on an ARMv7-a core with a 32 kB L1 cache and 64 B long cache-line, standard GCC compiler (v4.7.3). For each trace we identified up to two different modes of the application, based on the distance of memory accesses (cf. Figures 6.10 and 6.11).

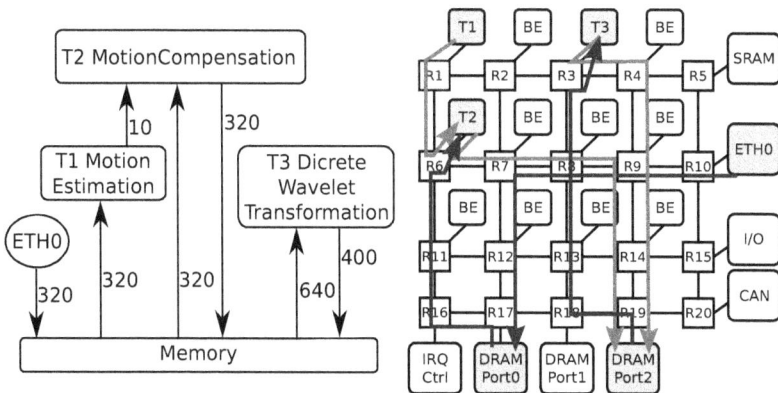

Figure 6.3: *Experimental setup with communication demands of SC tasks in MB/s for 1080p resolution and 40 fps (left) and mapping (right).*

Table 6.5: DDR port mapping for the use case.

Port	Reader	Writer
0	T1, T2, BE11, BE12	ETH0, BE11, BE12
1	BE2, BE7, BE8, BE9, BE13	BE2, BE7, BE8, BE9, BE13
2	T3, BE4, BE14	T2, T3, BE4, BE14

For the different approaches we followed baseline configuration schemes, without any optimizations. The overhead against the time budget was confirmed with the formal analysis from Section 3.4 done in the *PyCPA* framework [62] and hence permitted the usage of the control layer. In the case of the priority based (SP) approach, we always provide the SC traffic with a higher priority than BE traffic, so that BE traffic is preempted in the routers, as soon as there is any pending SC data stream. Following this scheme, the TIS approach avoids concurrent accesses of SC and BE senders to shared parts of the network. Hence, on an activation of a SC sender, its client will send a control message to all interfering BE nodes, blocking their network access. This is relaxed by the ALD approach. Consequently, instead of directly blocking BE senders, they are allowed to take the YX route to its destination, if this path is free. The SALDT approach accounts for different possible behaviours (runtime characteristics) of BE senders and allows a single BE sender (randomly selected during design time for the experiments) to concurrently share a path with SC traffic, as long as it stays below a certain threshold. For this, the client of a SC sender sends the requested throughput to each interfering BE node, which then calculates the remaining link bandwidth for obtaining the threshold value. With FALDT, the clients of the BE senders also exchange control messages with data about the current link usages (occupancy). This information is used for dynamic selection of BE senders, instead of static BE allocation as it was done in aforementioned approaches. This allows an improved load distribution for BE traffic and, therefore, higher utilization of the available network resources at runtime. Consequently, multiple BE senders are allowed to concurrently share the same path with SC traffic, as long as the accumulated interference stays below a threshold. And finally, for the SCC approach, the SC senders also receive the usage information of BE senders and dynamically decide, whether to use the XY or YX route.

6.2 Performance Results

Figure 6.4 presents the achieved results for a setup in which we use only two virtual channels (VCs) of the priority based NoC (cf. Section 5.3.2). The number of available VCs is restricted to two to increase the interference between the different traffic streams. For the static prioritzation (SP), the SC traffic was allocated on the high priority VC, whereas for the other approaches, SC and BE senders were sharing the low priority VC and the control layer was using the high priority VC. Hence, no additional hardware in the network is needed for the control layer as it uses the existing data layer to transmit the control messages. The figure shows *Tukey boxplots* of the normalized latencies for the different nodes in the network for accesses to the memory. For each run, the box covers 50 % of the latencies, with its lower and upper borders giving the 25 % and 75 % quartiles. The whiskers show the lowest (highest) value still within 1.5 of the inter-quartile range of the lower (upper) quartile. The median and average among the measured latencies are marked by a black bar and a red brick. For each of the nodes, we generated 100 random sets of interfering workloads and normalized the latency to the average latency when using SP.

It is visible, that TIS offers safety at the cost of decreased BE performance (i.e. an increased latency) compared to the simple prioritization of SC traffic. Only in a few cases (e.g. BE2 and BE8) it can slightly increase the performance of BE senders. However, TIS can be used to implement more priorities than available VCs. Hence, for systems with a restricted number of virtual channels, the prioritization approach (SP) might not be possible. With the allowance of alternate paths for BE traffic (ALD) we can reduce the negative impact of the control layer in most cases (e.g. BE9 and BE7), which strongly interfere with SC traffic. However, these alternate paths for BE traffic also lead to an increase of the latency for other BE senders, as for example BE4. This shows that the interdependencies between the different streams and the control layer must be better accounted for when configuring the approaches. As soon as we also account for the actual behaviour of BE traffic, we can further decrease the latencies for BE senders. For some nodes, e.g., BE2 and BE8, the additional information exchanged by BE senders that is used for the optimization of the accesses to the network can even lead to better results than the simple prioritization scheme.

For the same experiments, Figure 6.5 shows the results, when using two dedicated physical channels as described in Section 5.3. In this case, the control messages are sent on an own physical channel and thus induce no (high priority) interference in the network to the other traffic (cf. Section 5.3.3). Additionally, the data layer is only using a single (virtual) channel. In the figure latencies are normalized to the case of using SP and virtual channels and hence are in the same range as the results before. Compared to the case when using virtual channels, we see a similar behaviour. The simple approaches TIS and ALD increase the latency for BE traffic, while the other can decrease it. Consequently, the improvement in the performance of BE senders directly depends on the amount of information used by control layer while making decisions. We can also see, that the latencies of BE senders are a lower than in the previous case. This is due to the interference, which the control messages are inducing into the network, when using virtual channels and prioritization of control messages. Whenever these are sent on a VC with a higher priority than BE traffic, they will additionally block BE traffic and hence decrease the performance benefit in the designs using the control layer. Figure 6.6 shows the average number of control messages exchanged during the experiment. The figure presents the fraction of control messages compared to the overall number of exchanged messages for the different considered scenarios. In the figure, we can observe that the overhead is small (5–6.5 %) and depends on the actually applied version of the protocol. As expected, the control layer implementing FALDT and SCC protocols induce higher number of control messages. This is due to the fact, that in these approaches the BE nodes are also emitting control messages.

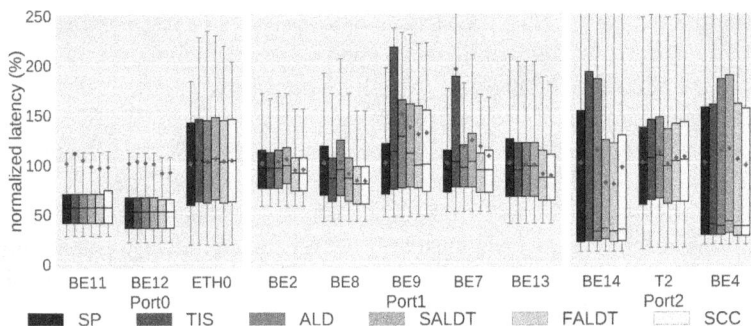

Figure 6.4: *Normalized end-to-end latencies using virtual channels.*

Figures 6.7 and 6.8 show the normalized latencies (including queuing delay at the source interface) for the BE senders *BE7* and *BE8* for various configurations of the network architecture for a single experiment (e.g. a sin-

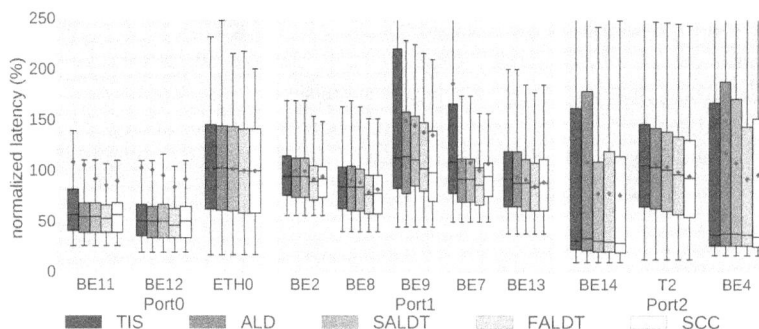

Figure 6.5: *Normalized end-to-end latencies using physical channels.*

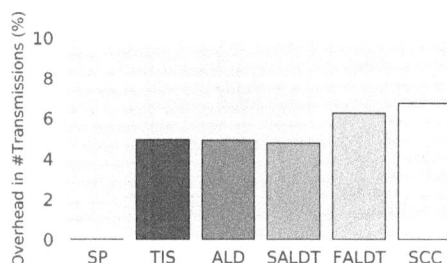

Figure 6.6: *Overhead of the control messages: percentage of control messages.*

gle random set of interfering workload). In the figures, the upper part (*write*) shows the latency from the computation node to the memory accounting for write and read requests. The bottom part (*read*) denotes the latency of the response (i.e. from the memory to the computation node). The figure compares different network configurations using static priority (SP) and the control layer with the temporal isolation (TIS) protocol and the selective congestion control (SCC). For the SP case, the SC senders use their own (higher priority) virtual channels whereas all BE traffic can use one, two or four VCs. Additionally, different sizes for the input queues of the routers are used with one, two, and eight flits as the buffer depth (denoted by Q in the figure). For the control layer approaches, a leading V denotes the use of a virtual control layer (i.e. sending the control messages on the highest priority

VC) and P denotes the use of a dedicated control NoC. For the control layer approaches, we differentiate the case where SC and BE senders use their own virtual channels or are sharing the same set of VCs. The former case is denoted as "x VC for BE" in the figure where x is the number of VCs that BE traffic can use. The latter is denoted as "x shared VCs" where SC and BE share x VCs.

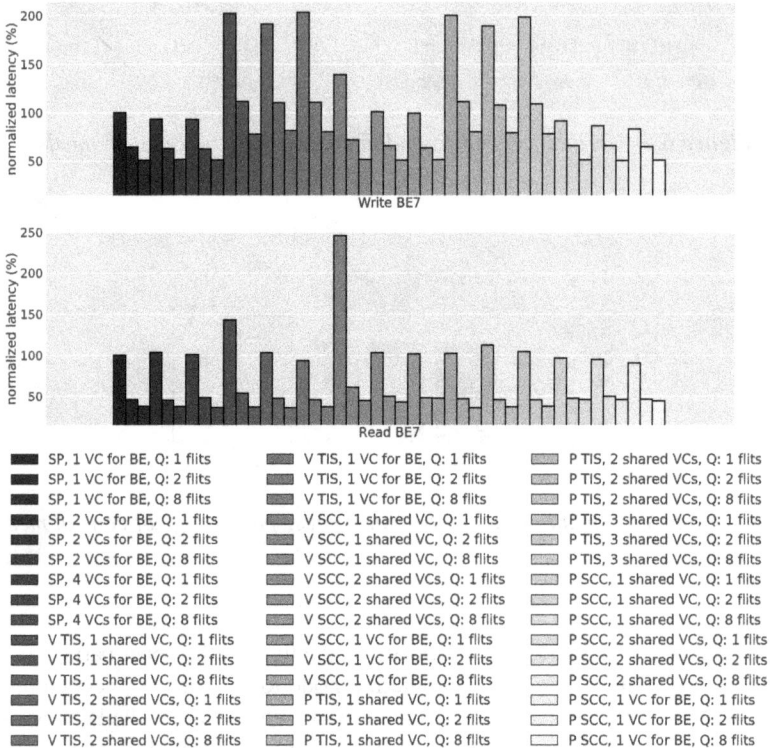

SP, 1 VC for BE, Q: 1 flits	V TIS, 1 VC for BE, Q: 1 flits	P TIS, 2 shared VCs, Q: 1 flits
SP, 1 VC for BE, Q: 2 flits	V TIS, 1 VC for BE, Q: 2 flits	P TIS, 2 shared VCs, Q: 2 flits
SP, 1 VC for BE, Q: 8 flits	V TIS, 1 VC for BE, Q: 8 flits	P TIS, 2 shared VCs, Q: 8 flits
SP, 2 VCs for BE, Q: 1 flits	V SCC, 1 shared VC, Q: 1 flits	P TIS, 3 shared VCs, Q: 1 flits
SP, 2 VCs for BE, Q: 2 flits	V SCC, 1 shared VC, Q: 2 flits	P TIS, 3 shared VCs, Q: 2 flits
SP, 2 VCs for BE, Q: 8 flits	V SCC, 1 shared VC, Q: 8 flits	P TIS, 3 shared VCs, Q: 8 flits
SP, 4 VCs for BE, Q: 1 flits	V SCC, 2 shared VCs, Q: 1 flits	P SCC, 1 shared VC, Q: 1 flits
SP, 4 VCs for BE, Q: 2 flits	V SCC, 2 shared VCs, Q: 2 flits	P SCC, 1 shared VC, Q: 2 flits
SP, 4 VCs for BE, Q: 8 flits	V SCC, 2 shared VCs, Q: 8 flits	P SCC, 1 shared VC, Q: 8 flits
V TIS, 1 shared VC, Q: 1 flits	V SCC, 1 VC for BE, Q: 1 flits	P SCC, 2 shared VCs, Q: 1 flits
V TIS, 1 shared VC, Q: 2 flits	V SCC, 1 VC for BE, Q: 2 flits	P SCC, 2 shared VCs, Q: 2 flits
V TIS, 1 shared VC, Q: 8 flits	V SCC, 1 VC for BE, Q: 8 flits	P SCC, 2 shared VCs, Q: 8 flits
V TIS, 2 shared VCs, Q: 1 flits	P TIS, 1 shared VC, Q: 1 flits	P SCC, 1 VC for BE, Q: 1 flits
V TIS, 2 shared VCs, Q: 2 flits	P TIS, 1 shared VC, Q: 2 flits	P SCC, 1 VC for BE, Q: 2 flits
V TIS, 2 shared VCs, Q: 8 flits	P TIS, 1 shared VC, Q: 8 flits	P SCC, 1 VC for BE, Q: 8 flits

Figure 6.7: *Normalized end-to-end latencies for BE7 for a single experiment and different network configurations.*

The figures confirm the results shown before. With an increased allowed dynamic, the approaches can improve the performance of the best-effort traffic. Another interesting observation is, that deeper buffers in the network routers are more effective (for the BE performance) than increasing the

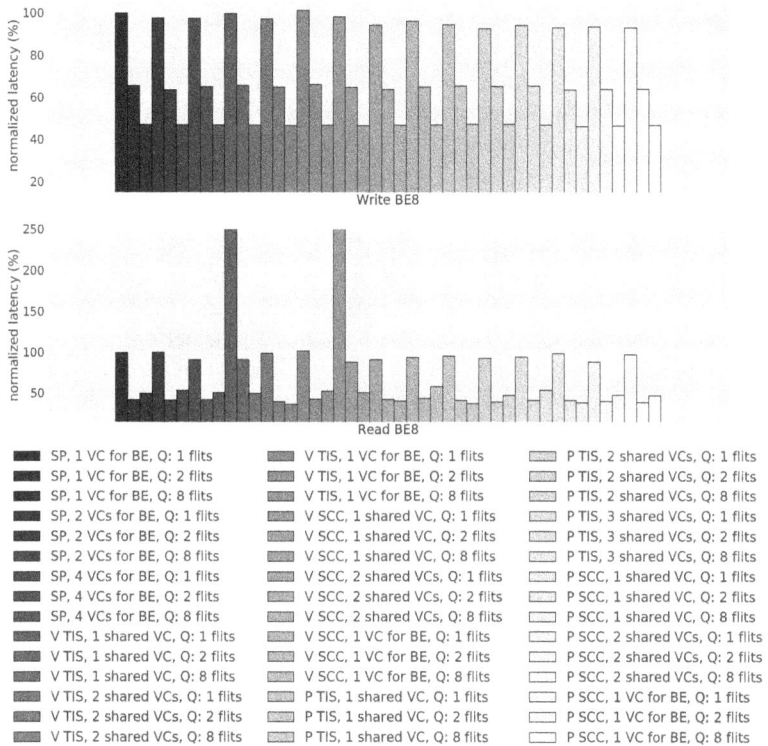

Figure 6.8: *Normalized end-to-end latencies for BE8 for a single experiment and different network configurations.*

number of virtual channels in the experiments. This effect results from the number of BE senders and interference at the traffic sources (e.g. the memory sending to different destinations). While an increased number of VCs can compensate for some interference, the available channels are quickly used up. As soon as no other channels are available, the flits of the packets get blocked. With small input buffers this blocking propagates from the congestion point up to the sender. Hence, the blocked packets blocks other traffic on multiple routers. With an increased buffer depth, it is more likely that a blocked packet fits into the network, such that not the whole path from the congestion point to the sender is blocked. This effect is especially important for the memory. If the memory tries to emit data to the network but the network channel is blocked, no further data can be emitted. The reason for this is, that the memory only has a single send queue. Thus, if a packet is blocked on a certain VC even traffic using another VC cannot be emitted. If the buffers in the network are sufficiently deep to buffer a packet that gets blocked in the network, the memory can still emit further packets using another VC.

Similar as shown before, TIS can lead to an increased latency for BE traffic. For the BE7 node, TIS increases in the write latency in all cases and the read latency in some cases. However, for BE8 TIS can decreases the write latency and for some cases also the read latency. This shows that even the conservative approach can help to improve the performance of the system in some cases. Still, the effect strongly depends on the behaviour of the network senders. The control layer protocols use the same configuration for all random sets of interfering workload. Hence, the protocols are not optimized towards the system including the BE workload. For a real system, the behaviour of the BE senders will influence the control protocol and the protocol (and the behaviour) will also affect the mapping of the BE senders. Following this, SCC shows a better performance than TIS and a similar or better performance than the static prioritization (SP). This is due to the fact, that SCC partly accounts for the actual behaviour of the network traffic. Even for cases where BE and SC senders are sharing a single virtual channel, the performance is comparable to the SP case, where SC and BE senders are using separate virtual channels. This shows that a global arbitration can help to efficiently share a virtual channel and thus to reduce the number of needed VCs. Still, the protocol and mapping are not optimized for the BE behaviour. Comparing the different implementations of the control layer, we see that using a dedicated control NoC can achieve a better performance than using a high priority VC for the control messages. Together with the ability to efficiently share a VC (e.g. to provide multiple priority levels and isolation on a single VC) and the increased adaptability of the system when using the control layer, a dedicated control NoC seems to be a reasonable choice.

Figure 6.9 presents the achieved results, where values consider both core execution and network communication. The figure shows the overall time of different benchmarks mapped to node BE8. This node is not sharing the memory port with any SC traffic but can experience indirect blocking. Again, the results are normalized to the case when using the prioritization scheme (SP). The results show that the TIS and ALD approaches decrease the performance for all benchmarks. This is due to the fact, that the control layer uses some safety margins when controlling BE traffic. Additionally, the control messages have some latency when traversing the network. Hence, instead of only being blocked by the transmissions of SC senders, some additional blocking through the protocol overhead occurs. When accounting for the actual load of the BE senders, the approaches can in average outperform the protocol overhead. However, the achieved performance depends on the use case and a few cases exists, where using the control layer is slower compared to the prioritization scheme. Only the SCC approach allows to improve the performance for all benchmarks. It allows to decrease the overall execution time by up to 30 %. The improvements of the adaptive approaches, compared for example to TIS, results from the different behaviour of the BE senders (e.g. varying network load) that can be accounted for. Figures 6.10 and 6.11 show such behaviour for two exemplary benchmarks. The figures show the distance between accesses to the network for 5000 accesses (read or write) to the memory. As can be seen, there are areas where several consecutive accesses have a high distance between each other (e.g. low load) and areas where only a low distance (e.g. high load) exists. For static approaches, the worst-case behaviour (e.g. assuming always the high load) must be accounted for, leading to pessimism and, thus, reduced performance. Dynamic approaches, on the other hand, can account for the low load durations and allow consecutive accesses of a (low load) BE sender and safety-critical traffic.

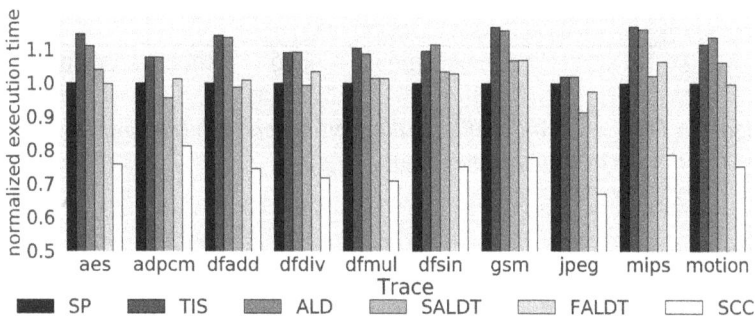

Figure 6.9: *Normalized execution time for various traces.*

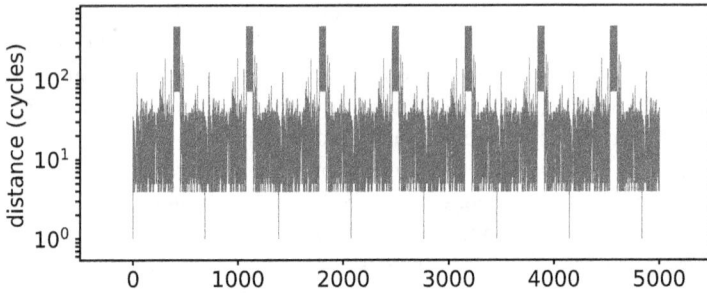

Figure 6.10: *Execution profile of the* motion *benchmark (distance between consecutive accesses to the memory).*

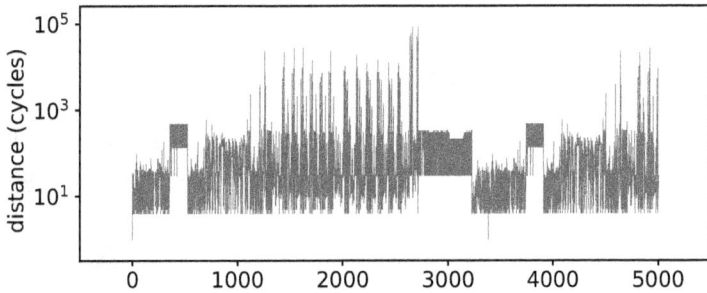

Figure 6.11: *Execution profile of the* jpeg *benchmark (distance between consecutive accesses to the memory).*

6.3 Synthesis Results

This section presents the hardware overhead of the control layer. The evaluation of the hardware overhead was done using the IDAMC platform [216] on a Virtex-6-LX760 Xilinx FPGA using *Xilinx ISE 14.6* with default optimization setting and no special optimizations for the *VHDL* implementation. The device utilization data were collected from the *Module Level Utilization Summary Report* produced by ISE. Tables 6.6 and 6.7 show the results. For the synthesis, we assumed for all approaches the same size of the stored tables (64 entries) and implemented the control logic in hardware. The first table shows the overhead for the network interfaces when using a distributed

control layer from Section 5.3.4. An evaluation of centralized control layer
are presented in [122; 230]. Table 6.6 shows that the simple approaches (TIS,
ALD, SALDT) introduces less than 5% overhead to the area of the network
interface (NI) module. The more complex approaches (FALDT, SCC) introduce slightly more overhead (5.5%). The slight increase for FALDT and
SCC compared results from the additional information and control logic that
is needed to estimate the optimal path. However, the approaches also allow to
account for more priority levels than are available by the architecture. Hence,
when needing more than two priorities, the overhead for the approaches will
stay the same, where for the classic prioritization one will have to increase
the routers (e.g. add more buffer space for virtual channels). When using
a centralized approach, the network interface extensions show a similar increase of the area (e.g. approximately 120 LUTs). Additionally, there is an
overhead due to the resource manager (RM). This varies between 230 and
1500 LUTs depending on the implementation [122; 230].

Table 6.6: *Synthesis results of a network interface on a Virtex-6 LX760
FPGA.*

Unit	SP	TIS	ALD	SALDT	FALDT	SCC
#Regs	2581	2702	2707	2715	2723	2723
#LUTs	4925	5160	5160	5162	5190	5192

Table 6.7 compares the area of different network routers. It compares
a basline NoC with four virtual channels against two approaches using the
control layer. For this, we implemented the control layer as a high priority
VC (5 VCs) and as a dedicated network (4 VCs+C-NoC) as introduced in
Section 5.3. Compared to the reference router (4 VCs) an additional virtual
channel (VC) (5 VCs) requires approximately 28% more register and 30%
more LUTs in the VHDL implementation. Whereas using a dedicated control
NoC for the control layer only requires 22% more registers and 18% more
LUTs. Hence, besides the performance and analysis benefits, a dedicated
network for the control layer seems to be reasonable. However, the table
focuses on the control logic of a network router. Hence, when using the
dedicated control NoC, there will be an additional area overhead due to the
increased wiring of the additional network routers. On the other hand, the
needed area for the memory might decrease, as the control layer allows using
smaller buffers (e.g. no additional interference through high priority control

Table 6.7: Synthesis results of network routers on a Virtex-6 LX760 FPGA.

Unit	4 VCs	5 VCs	4 VCs+C-NoC
#Registers	1124	1433	1366
#LUTs	1416	1841	1661

messages and avoiding of blocking propagation through the control layer instead of deep buffers).

6.4 Evaluation Against Requirements

Section 1.4 discussed the requirements of (future) safety-critical embedded systems while Section 5.2 listed requirements for the implementation of the control layer of the NoC-RM. In the following we discuss, how the proposed architecture with the control layer satisfies these requirements.

The requirements from Section 1.4 can be broken down into three major areas: *efficient support of different traffic types*, *low cost*, and *flexibility*. However, these are tightly coupled and influence each other.

The efficient support of different traffic types covers several aspects. First, the system shall allow an efficient co-hosting of applications with different requirements. That is, for example, providing formal worst-case guarantees for safety-critical (real-time) traffic, while at the same time high performance for non safety-critical traffic. And secondly, the system needs to provide sufficient means to prevent faults (cf. Table 1.1) to influence the functioning of the system. This especially includes fault isolation between the non safety-critical and safety-critical parts of the system. For NoCs, both can be summarized as *sufficient independence* between traffic classes or streams. All these aspects are solved by the proposed NoC architecture using the NoC-RM. The NIs with the client/control extensions allow to isolate network nodes and traffic streams from each other. This includes timing interference (e.g. rate limiting) and access protection (e.g. address translation tables). Interfering safety-critical and non safety-critical senders are synchronized using the NoC-RM. This enables to isolate traffic streams and also to use a work-conserving, dynamic arbitration and thus to give the BE senders access to the NoC, whenever there is sufficient slack available for the safety-critical senders. And hence, this allows to improve the performance of BE traffic as shown in Section 6.2. The client extensions in the NI provide a fine-granular support for different admission control mechanisms (rate limiters, address translation tables) and thus enable to efficiently handle different transmission kinds (e.g. short cache-based messages and long DMA transfers). However, not all transmissions are required to use the synchronization and, moreover, the NoC-RM can completely be disabled, allowing to use the underlying network *as it is*.

These general quality of service (QoS) related properties are extended by standard means for error detection like a CRC check for flits, a header parity check, a sender ID (implicitly encoded in the route), a firewalling mechanism, and a sequence counter in the flit header. However, as the focus of this work was on QoS, the error handling support is only rudimentary in the current design. With these mechanisms, all errors from Table 1.1 can be detected as indicated in Table 1.2. While no unique data ID is present, the *address translation*, *sender ID*, and *firewall* can be used to prevent an incorrect addressing. Additionally, the use of a dedicated control network allows a fine-granular error detection and handling. As the NoC-RM can use the control network to accumulate error reports from the network interfaces and network routers, it enables a global error view and handling. This global view helps to more efficiently route traffic around detected faulty network parts and to provide the system software with more accurate error reports.

The costs cover two major areas: the hardware costs/overhead and the design or development costs of a system. From the hardware cost perspective, the NoC-RM provides some benefits. It can be implemented in software (if NoC provides priorities). This can deliver better properties, such as safety, performance, or flexibility while not needing additional hardware [122; 125]. Moreover, it can be used to decrease the needed buffer space (e.g. less VCs) in the NoC as contention and thus blocking propagation can be avoided. However, the use of a hardware extensions (e.g. client modules in the NI and a hardware accelerated RM) and a dedicated network for the control messages can improve the benefits (cf. Section 6.2). Hence, the proposed architecture uses additional hardware components inducing some overhead. However, most QoS mechanisms, especially when they also provide performance for non safety-critical components, induce hardware overhead. In contrast to other QoS extensions, the proposed architecture allows to mitigate the head-of-line blocking and blocking propagation problems. Hence, the buffer sizes and the number of virtual channels in the network can be reduced. In such a way, the additional hardware of the NoC-RM helps to reduce the size of the data layer, limiting the overall hardware overhead. Similar to this, the ability of the NoC-RM, to provide more priorities than available virtual channels helps to decrease buffer size.

Another important cost aspect of a system are the design and verification efforts. The proposed architecture uses the NI (with address translation, firewall, and rate limters) to isolate nodes from the network and each other. Additionally, the Noc-RM (together with the NI) offers full control on the network traffic. This enables a contract/interface based design and helps to analyse the system (as worst-case traffic behaviour is known/enforced). Hence, it simplifies system design and verification. Furthermore, the use of a dedicated control network allows to design and verify the control layer of the NoC-RM independently of the data layer. Hence, it can be re-used without costly re-design and re-verification (of the control layer). At the same time, this allows to re-use existing data layers and there is no need to re-design the whole interconnect. This is also the case, when a different resource allocation scheme should be used. In that case, there is no need to adapt the network hardware as the allocation is controlled by the NoC-RM.

The possibility to switch between different resource allocation schemes also enables a high flexibility of the design. As discussed, switching between different resource allocation schemes is relatively simple with the NoC-RM. It requires only reprogramming of the central resource manager or the control modules in the NIs. This even allows to use different schemes for different parts of the system or groups of senders. For instance, for selected nodes a static priority based scheduling can be applied, whereas for other regions strict temporal isolation of senders with a TDM based scheduling can be applied. Additionally, the switching between different schemes can be done at run-time or offline. Hence, a running system can adapt to changes (e.g. varying environmental conditions, different systems modes, or updates) or the same design be applied in different use cases that need different resource allocation schemes. That is, the application of different resource allocation schemes does not require modifications of the (data layer) routers. Therefore, it is possible to offer one chip with a generic network architecture and hence offer safety and real-time as a feature for a specific design.

To efficiently achieve these properties, Section 5.2 required a *low latency of control messages*, a *low interference of the control messages on existing traffic*, and a *low hardware overhead*. This is solved by using a dedicated control network. The control messages should not be blocked by any other traffic to permit a low latency for the control layer. While this allows to use a high priority virtual channel, this approach would induce additional traffic to the network that interferes with the normal traffic (cf. Section 6.2). Hence, the proposed architecture uses a dedicated network solely for the control messages. As there is no other traffic in this network and the router architecture is optimized for the control messages, they achieve a low latency. Additionally, they induce no additional interference to the data layer. This approach introduces a second network to the system and hence hardware overhead. However, all QoS mechanisms introduce additional hardware overhead to the system. As the NoC-RM allows to better utilize the existing hardware, as it, for example, avoids the propagation of blocking, and offers the needed flexibility, it seems feasible to use the NoC-RM. Hence, we can focus on the hardware overhead of the different implementations of the control layer of the NoC-RM. Comparing the two main possibilities (high priority virtual channel and dedicated network), the overhead is similar. When using a high priority VC, additional buffers are needed but the router itself can be re-used. However, as the data layer typically provides a bigger payload and more features than needed, there is some over-provisioning for the control messages. When using a dedicated control network, on the contrary, the network can be optimized to the control messages. That is, smaller packet and flit sizes and thus smaller buffers can be used.

With the discussion above, we can summarize the properties of different QoS mechanisms and a basic performance oriented NoC using round-robin (RR) arbitration as shown in Table 6.8. The table breaks down the requirements to eight different aspects. *Performance* indicates whether the network provides a low latency for best-effort streams and if it is work-conserving. The *hardware overhead* was discussed above. It summarizes how much additional hardware is needed, compared to a simple round-robin network. The *design effort* denotes, if the design is simple or if inter-dependencies between components challenge the correct design. *Scalability* combines the first three aspects and reasons on how they challenge the scalability. That is, a bad scalability results is higher cost, lower performance or higher design effort. *Flexibility* covers the re-useability and adaptability of the system as discussed above. The *verification overhead* denotes the additional effort to verify a system using the mechanisms. Typically, static approaches with well-defined application behaviours can be verified more easily than highly dynamic systems. Along with this, *safety/predictability* denotes the general possibility to derive (tight) upper bounds on the behaviour. In general, it is possible to derive upper bounds for all mechanisms. However, when no QoS is used at all (e.g. in the case of a non-preemptive round-robin), the analysis results get very pessimistic. And similar to this, *analysis effort* represents the easiness to analyse the system. Typically, a lower number of dynamics and well-defined traffic behaviours simplify the analysis.

The NoC-RM approaches can satisfy the most requirements in the table. For the NoC-RM just one to three requirements cannot fully be satisfied. When using a virtual control layer, the whole control logic can be implemented in software. That way, there is a very low hardware overhead. However, if hardware extensions are used to improve the performance of the control layer, it induces an additional overhead. Along with this, the design effort changes. While a software design is still complex, it is a re-useable software and can easily be adapted. When using hardware extensions, these must be integrated into the design, leading to a higher effort. Additionally, as the control messages are sent on a high priority VC, they interfere with other traffic. Hence, this additional overhead must be accounted for, when designing the system. The latter also challenges the *verification overhead*. The chosen protocol, or more precisely the amount of sent control messages, now induce an additional load to the system. This can lead to a dependency during the analysis and the design, challenging the verification. Still, this factor can be accounted for. When using a dedicated control layer, just the *hardware overhead* is a slight drawback of the approach (as for many other QoS approaches). This is due to the fact, that the network interfaces must be extended and an additional interconnect is used. However, as this interconnect is isolated from the remainder of the system, it helps to reduce the design effort and verification overhead. Overall, a dedicated interconnect for the control layer is a reasonable choice to build a system that provides QoS for safety-critical traffic, high performance for non safety-critical traffic, and flexibility at the same time.

Table 6.8: *Properties of Different QoS-Mechanisms (√ : good; − : fair; × : poor).*

	Performance NoC (e.g. RR)	Spatial Isolation	Static Temporal Isolation (e.g. TDM)	Dynamic Temporal Isolation (e.g. SPP)	Virtual Control Layer	Physical Control Layer
Performance	√	√	×	√	√	√
Hardware Overhead	√	×	√	−	√ / −	−
Design Effort	√	√	√	√	√ / −	√
Scalability	−	×	×	×	√	√
Flexibility	−	×	×	−	√	√
Verification Overhead	×	√	√	×	−	√
Safety/ Predictability	×	√	√	√	√	√
Analysis Effort	×	√	√	−	√	√

6.5 Summary

In this chapter we evaluated the architecture supporting the NoC-RM presented in Chapter 5. For this, we compared different protocols of the NoC-RM against the commonly used prioritization of safety-critical traffic. This comparison includes the implementation of the control layer of the NoC-RM as a dedicated physical channel and on a high priority virtual channel.

Using the NoC-RM, we were able to increase the performance of non safety-critical senders by up to 30 % using the SCC approach without endangering the timing of safety-critical senders requiring real-time guarantees. These findings were confirmed by experimental evaluation. Hence, the NoC-RM allows similar performance benefits as the presented NoC router extensions from Sections 4.3 and 4.4. At the same time, the NoC-RM approach induces less than 5.5 % hardware overhead at the network interfaces.

Using an additional physical interconnect for the control layer of the NoC-RM achieved better performance results than using a high priority virtual channel. This is due to the additional blocking the control messages induce to the system, when using a virtual control layer and sending the control messages on a high priority VC.

Moreover, the results showed a strong dependency on the configuration of the individual approaches and the workload induced by the network nodes. Hence, to further exploit the benefits of the NoC-RM, more sophisticated approaches to select the parameters should be used. Additionally, the system design and mapping phases need to account for the NoC-RM to further optimize the results.

In summary, the architecture satisfies the requirements of future mixed-critical systems that execute safety-critical and non safety-critical real-time application. Hence, an architecture using the NoC-RM for providing QoS and a dedicated interconnect for the control layer provides a feasible and appealing alternative for the future embedded systems requiring high performance and real-time guarantees.

7. Conclusion

The industry of safety-critical and dependable embedded systems calls for even cheaper, high performance platforms that allow flexibility and an efficient verification of safety and real-time requirements. In this sense, flexibility denotes the ability to (online) adapt a system to changes (e.g. changing environment, application dynamics, errors) and the reuse-ability for different use cases. To cope with the increasing complexity of interconnected functions and to reduce the cost and power consumption of the system, multicore systems are used to efficiently integrate different processing units in the same chip. Networks-on-chip (NoCs), as a modular interconnect, are used as a promising solution for multiprocessor systems on chip (MPSoCs), due to their scalability and performance. Hence, future NoC designs must face these challenges. Traditional NoC concepts address only a sub-part of the challenges—they focus on either performance or predictability. And existing, predictable NoCs are deemed too expensive and inflexible to host a variety of applications with opposing constraints. In this work we tackled the challenges and developed a predictable and runtime-adaptable NoC architecture that efficiently integrates mixed-critical applications with opposing constraints.

In the first part of this thesis, we presented design challenges in the domain of future safety-critical and dependable embedded systems. One major aspect for these systems is the verification of the timing behaviour. This is tackled by the use of formal analysis frameworks. However, state-of-the-art analysis frameworks for NoCs make some simplifications when analysing a system. One is to assume that buffers never overflow (e.g. that backpressure does not occur). While this can be guaranteed by the use of large buffers in the network or by using very restrictive rate limiters (i.e. allowing only a very low load), these render such analysis approaches unfeasible for real-world systems. In embedded systems the buffer size is strictly limited (to save cost) and the application should use the network with the highest possible performance in order so avoid over-provisiong of resources. To solve this problem, we presented a new analysis approach based on the compositional performance analysis (CPA) in Chapter 3 that can handle backpressure. With this analysis, realistic embedded systems with limited buffer space can be analysed. This enables to use smaller buffers when designing a system, as no over-provisioning is necessary to render the analysis possible.

However, the results of the evaluation of the analysis (cf. Section 3.5) demonstrated that backpressure and blocking propagation can lead to overly pessimistic results, especially for systems with shared buffers. While quality of service (QoS) mechanisms can be used to reduce the adverse impacts, e.g. through limiting the conservatism used in the analysis, there is also the possibility to improve the analysis. Hence, future work should focus on the reduction of the conservatism used in the proposed analysis. One example for this is the adaptation of the backpressure blocking. In the current analysis, the backpressure blocking is used as an individual blocking factor that also takes part in various other blocking factors. But, typically the backpressure blocking on a port exhibits a sub-additive behaviour. That is, the backpressure blocking accounting for four events is smaller compared to summing up the backpressure blocking of three and a single event. Hence, in future work a proof for this assumption can be derived and used to tighten the analysis results. For this, the backpressure blocking could be calculated per port instead of per stream, which would require to first derive the accumulated event model to each port and then to derive the blocking.

In the second part of this thesis, we proposed and evaluated different QoS mechanisms for modern NoC architectures for mixed-critical real-time systems. For this, we started in Chapter 4 with an overview on existing QoS mechanisms for NoCs. Most of the existing solutions are not capable of solving the requirements of future mixed-critical system (cf. Chapter 1). Typically, these solutions sacrifice system performance (especially for the non safety-critical parts) to achieve a predictable behaviour for the safety-critical applications. To solve this problem, we developed two hardware based QoS extensions for NoC routers in Sections 4.3 and 4.4. We showed that these can provide the required predictability and high performance for the non safety-critical parts at the same time. However, as the use of hardware based router extensions limits the flexibility, we also investigated a HW/SW co-design approach that uses an end-to-end synchronization of resource accesses: the network-on-chip resource management (NoC-RM). This solution uses a control layer to apply a global resource scheduling and network admission control (cf. Section 4.5). The NoC-RM was rendered as a feasible choice for future mixed-critical systems. Hence, we developed an architecture that provides support for the control layer of the NoC-RM to increase its benefits in Chapter 5. We first derived the requirements for a NoC architecture that uses and supports the control layer. This resulted in two main possibilities to implement the control layer: as a high priority virtual channel or as a dedicated control network. In booth cases, the control layer can be used to provide QoS while at the same time high performance for the non safety-critical parts. Additionally, as the NoC-RM enables to mitigate the problem of blocking propagation, it allows to decrease the buffer sizes of the (data) network. To prove the benefits of the architecture, we conducted an experimental evaluation and compared the two possibilites against the commonly used prioritization of safety-critical traffic in Chapter 6. For this, we used different protocols for the NoC-RM. The simulation showed that using an additional physical interconnect for the control layer achieved better performance results than using a high priority virtual channel. In both cases, the dynamic arbitration of the NoC-RM can achieve better performance that the classic prioritization. In summary, the evaluation showed that the architecture satisfies the requirements of future mixed-critical systems that execute safety-critical and non safety-critical real-time application. Hence, an architecture using the NoC-RM for providing QoS and a dedicated interconnect for the control layer provide a feasible and appealing alternative for the future embedded systems requiring high performance and real-time guarantees.

There are still multiple open questions that provide a direction for future research regarding the NoC-RM and its implementation. First, the results of the evaluation showed a strong dependency on the configuration of the individual protocols of the NoC-RM and the workload induced by the network nodes. Hence, to further exploit the benefits of the NoC-RM, more sophisticated approaches to select the parameters should be used. Additionally, the system design and mapping phases should account for the control layer to further optimize the results. These topics need further investigation to provide a proper design methodology for a system using the control layer.

Next to this, this thesis (and the work related to the NoC-RM [122; 125; 223]) focussed on QoS, e.g., on controlling the interference. However, using an additional interconnect to monitor and control the remainder of the system, offers several other possibilities. Some concepts extending the presented NoC-RM will be discussed in the following.

7.1 Concepts Extending the QoS Control Layer

This section presents some concepts extending the NoC-RM to enable a sophisticated self-aware NoC/system control. These concepts can be used as an extension to available architectural mechanisms (e.g. to increase the benefits) or to replace them by an end-to-end approach and thus possibly reduce the complexity of the interconnect (e.g. error handling through RM not in each router). However, these topics were not in focus of the research activities during this thesis and just highlight some possibilities on a high level of abstraction for future research directions. And while it seems feasible to extend the NoC-RM with certain features, this section does not imply that the extended NoC-RM shall do this using the presented control layer architecture. That is, when more features are added to the NoC-RM, its complexity and possibly its timing/performance impact may increase. Hence, if many features are needed that shall be not implemented in the data layer but in the NoC-RM, an architecture supporting multiple control layers (virtual or physical) might be the better choice. In such a design, each control layer will implement a certain set of functionalities to support the data layer or the system at all.

7.1.1 Quality of Service in the Data Layer

When developing the NoC-RM and its network architecture, the main goal was to introduce QoS, performance, and flexibility to existing architectures with only small changes to existing architectures. Hence, the NoC-RM assumed a simple NoC with priority based arbitration for the data layer. While such an architecture (together with the control layer) is already capable of satisfying all requirements, it might not be the best solution for every use case.

Chapter 4 presented QoS-aware router architectures, that are promising with respect to their isolation and performance properties. Their main drawback was the flexibility. One possible future research direction is the combination of a data layer based on such an architecture and the NoC-RM. These mechanisms use parameters to configure the actual network behaviour. Examples are the blocking counters, the threshold values for the buffers, or the available (dynamic) priorities used. In general, the NoC-RM can be used to adapt these parameters during the run-time of the system according to the global state of the system. This would introduce flexibility to the system while possibly also increasing the benefits of these mechanisms. Additionally, other mechanisms, such as *Virtual Networks*, which enables a region based QoS approach, exist [95]. Hence, future research might focus on the interaction of the NoC-RM with other QoS mechanisms to find an optimal architecture.

7.1.2 Monitoring

So far the NoC-RM uses the client interfaces in the network interfaces (NIs) to detect and manage accesses to the network, as well as simple error checks in the routers and NIs. To increase the capabilities of the system to adapt to unpredictable changes (cf. autonomic computing [112; 165]), this input to the NoC-RM can be extended to use various monitoring information. One benefit of doing such monitoring on an additional layer is that the monitoring data itself does not interfere with the traffic on the data layer. Such interference can distort the actual measurement of the data layer. Examples for additional monitoring information that can be used are the current load/utilization of a link/port or virtual channel, the temperature, the long term load, or other ageing affecting factors. These would amplify the possibility of the system to adapt to changes in the behaviour and environmental conditions. For example, if the temperature of the environment or a network router increases, the affected router can be slowed down or traffic re-routed to relieve the router. Besides these simple monitors, even more advanced monitors are possible, that observe the actual behaviour of an application or traffic streams. With this, the NoC-RM could identify suspicious traffic patterns that indicate side channel attacks. Extending the simple request-based protocol by monitoring information thus enable a context-aware self-management as sketched in Figure 7.1. In such a system, the designer (or programmer) provides policies and specifications (e.g. for requirements and desired system behaviour). The NoC-RM uses the available sensors (and arriving requests) to collect the context information. Based on this and the informations from a repository, which contains the context, policy, and requirements provided by the designer or user, a view on the current system can be obtained. From this, a new configuration (e.g. system state) is estimated, that adheres to the given policies and requirements. And with the known new configuration, the NoC-RM enforces the new behaviour (e.g. through re-configuration of rate limiters or adapting traffic routes).

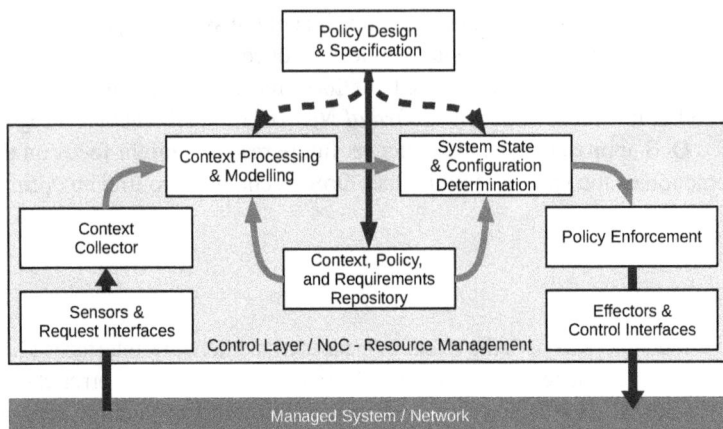

Figure 7.1: *Context-aware self-management of a NoC.*

7.1.3 Power

The NoC-RM can be used to optimize the power consumption of the network. Power saving techniques, as for example power gating of network routers, use local information at a router to decide if it can be turned off. However, as there is no knowledge about when new data will arrive, this approach can lead to cases in which the router is switched *off* and *on* too frequently diminishing the power saving (as there is power overhead for doing this) or in which data has to wait at an offline router until it is turned on. In state-of-the-art mechanisms, this is tackled by look-ahead mechanisms. Here, additional logic and wires are used, to inform routers multiple hops ahead about the arrival of new data. While this solves the problem of additional waiting, it introduces hardware overhead. Other approaches dynamically adapt the frequency of the network routers to keep the balance between energy-saving and performance loss. Such approach, for example, use power management units in the network routers that collect monitoring information and intelligently control the power shutdown mechanisms and frequency [79].

The NoC-RM has a global knowledge about the traffic and is informed about a new network connection, before a packet enters the network. Hence, the NoC-RM can tun on routers in advance. That way, the data flits observe no additional delay as the router are always online, when the flits arrive. Additionally, the NoC-RM can be used to re-route traffic. This enables to account for the power state of network paths to optimally distribute the packets for lower power consumptions. The same approach can be used to guarantee that the thermal stress and ageing of the network routers is equally.

With an existing dedicated control layer it is also possible to completely shutdown the routers of the data layer. In state-of-the-art solutions, some parts of the routers (e.g. the logic to detect new data and the look-ahead mechanisms to wake up routers early) need to stay online. This is not necessary when using a dedicated control layer. However, the control layer itself induces some (power) overhead and needs to stay online. While the control layer itself can use power saving mechanisms, it will experience the same drawbacks as the data layer. But as the complexity of the control layer is typically lower than the data layer, it might still be more power efficient. And while it is not clear whether it is feasible to introduce the control layer solely for the power management, it seems feasible to re-use an existing control layer for this purpose.

7.1.4 Errors

The additional network and the global knowledge of the control layer can be used to optimize the error detection and handling. An example are transient errors in the network. These can have permanent effects (e.g. reserving a wrong port that is never released again). To solve this, routers need additional logic to detect and prevent these errors [181; 182]. Instead of this additional logic, the control layer can use its global view to detect such errors and reset the affected routers. This global view also helps to identify corrupted paths of the network or parts of the system and to re-route traffic around the affected nodes. Hence, the error checks in the routers and the NI can be extended with end-to-end checks, distributed invariance checkers similar to the *ForEVeR* framework [167; 168] or *NoCAlert* [51]. The global management of the communication resources also helps sending test packets or patterns without endangering other ongoing traffic. Such test packets can be used to detect faulty NoC regions [98]. However, these packets should not interfere with safety-critical real-time traffic and hence only be sent when the network (or the path under test) is free.

The global knowledge of the NoC-RM also helps to solve another challenge of local error handling methods, especially for safety-critical traffic. Recovery (or reset or rollback) mechanisms typically induce timing overhead, as traffic needs to be resent. This must be accounted for, when designing the system, which typically leads to conservative assumptions and over provisioning. That is, there is some additional load for each component resulting from the error recovery that must be accounted for. The NoC-RM can be used to reduce the negative (timing) impact in such cases. This can be done by applying the recovery only when there is no safety critical traffic (assuming the error only affects BE senders so far), to re-route safety critical traffic to minimize the performance impact of the rollback, or to re-configure the QoS parameters.

7.1.5 Debug and Testing

The NoC-RM can be used to help in debugging and testing of a NoC based system and of the NoC itself. A problem of debugging a system, especially during post-silicon tests, is the restricted observability of internal signals and the time of issuing test transactions and their coverage of possible NoC and router states. As a result, it is difficult to detect and debug a failure when a test fails. Additionally, when the test succeeds the verification engineer has limited knowledge on whether the test really affected the desired parts of the system. For example, if a NoC uses an adaptive routing to mitigate link or router errors, the verification engineer needs to check all possible paths and check if the system is triggering its online fault tolerance mechanisms correctly. To help the verification engineer, systems typically add support to the design of a chip (e.g. design for debug (DfD) [57]).

A control layer can be used to increase the observability of the network and also to provide (and enable) a *debug mode* in a router. In such a case, the control layer can monitor the actual behaviour of the router and network links, as well as inject test patterns (or trigger their generation). Together with the end-to-end monitoring of the NIs, this enables an in-deep view of the system functioning. Similar approaches are already in the focus of current research. For example, Vermeulen and Goossens [228] present a monitoring infrastructure that can be used for performance analysis and debug. Goossens et al. [82; 83] implement a debug architecture that is used to supplement conventional computation-centric debug with communication-centric debug methods. Tang and Xu [208] use debug probes between nodes (e.g. cores) and the NIs that hep to debug the node and also support inter-core transaction analysis. Friederich et al. [78] propose debug probes within each network router and monitoring units to trace the activity of each node. And Stępniewska et al. [203] use real-time packet sniffing and monitoring. This approach uses a dedicated debug mode in which each packet is additionally sent to an off-chip control device. The mentioned approaches can be combined with the NoC-RM. Hence, future research might focus on an efficient design for debug support of the NoC-RM to ease the verification of the system.

A. Appendix: Publications

A.1 Related Publications

The following papers, based on the work presented in this thesis, have been published:

A.1.1 Reviewed

[216]: S. Tobuschat, P. Axer, R. Ernst, and J. Diemer. "IDAMC: A NoC for mixed criticality systems". In: *2013 IEEE 19th International Conference on Embedded and Real-Time Computing Systems and Applications*. Aug. 2013, pages 149–156. DOI: 10.1109/RTCSA.2013.6732214
This work explains initial Integrated Dependable Architecture for Many Cores (IDAMC). The IDAMC provide spatial and temporal isolation and thus sufficient independence. This is achieved with the use of different monitoring and control mechanisms in the network interfaces and the network providing guaranteed service through the *Back Suction* scheme. This architecture was used as a baseline architecture and extended during this thesis to incorporate the control layer. Chapter 5 is based on this work.

[217]: S. Tobuschat, M. Neukirchner, L. Ecco, and R. Ernst. "Workload-aware shaping of shared resource accesses in mixed-criticality systems". In: *Hardware/Software Codesign and System Synthesis (CODES+ISSS), 2014 International Conference on*. Oct. 2014, pages 1–10. DOI: 10.1145/2656075.2656105

This work presents the general idea of a workload aware shaping that can be used in network interfaces or routers. Unlike many other existing approaches, it prioritizes best-effort traffic whenever possible and use an accurate acquisition of the workload. With this it achieves an improved performance for general-purpose functions. At the same time it provides full guarantees to safety critical real-time functions. Hence, this work explains the basic of using the slack of safety-critical applications to improve the performance for non safety-critical applications. Chapter 3 is partially based on this work.

[220]: S. Tobuschat and R. Ernst. "Real-time communication analysis for Networks-on-Chip with backpressure". In: *Design, Automation Test in Europe Conference Exhibition (DATE), 2017*. Mar. 2017, pages 590–595. DOI: 10.23919/DATE.2017.7927055
This work introduces a real-time communication analysis for best-effort networks-on-chip with finite sized buffers and backpressure. The analysis allows exploiting the behaviour of individual traffic streams to determine safe upper bounds on the latency of individual packets. It was the first safe NoC analysis that can handle backpressure. Chapter 3 is based on this work.

[219]: S. Tobuschat and R. Ernst. "Efficient Latency Guarantees for Mixed-Criticality Networks-on-Chip". In: *2017 IEEE Real-Time and Embedded Technology and Applications Symposium (RTAS)*. Apr. 2017, pages 113–122. DOI: 10.1109/RTAS.2017.31
This work presents a novel arbitration scheme for NoC routers in mixed-criticality systems. Unlike many other existing approaches, it prioritizes best-effort traffic over safety-critical guaranteed-latency traffic whenever possible. Thus, it exploits the latency slack of critical applications to improve the performance for general purpose functions in mixed-criticality systems. Chapter 4 is based on this work.

[221]: S. Tobuschat and R. Ernst. "Providing Throughput Guarantees in Mixed-criticality Networks-on-Chip". In: *2017 30th IEEE International System-on-Chip Conference (SOCC) (SOCC 2017)*. Munich, Germany, Sept. 2017, pages 207–212—best paper award
This work presents a novel arbitration scheme for NoC routers in mixed-criticality systems. Unlike many other existing approaches, it prioritizes best-effort traffic over safety-critical guaranteed-throughput traffic whenever possible. Thus, it exploits the latency slack of critical applications to improve the performance for general purpose functions in mixed-criticality systems. Chapter 4 is based on this work.

[223]: S. Tobuschat, A. Kostrzewa, and R. Ernst. "Selective congestion control for mixed-critical networks-on-chip". In: *Integration, the {VLSI} Journal* (2017), pages -. ISSN: 0167-9260. DOI: `https://doi.org/10.1016/j.vlsi.2017.12.003`. URL: `https://www.sciencedirect.com/science/article/pii/S0167926017302341`
This work presents an online management approach for NoC resource. For this, it proposes a safe mechanism to exploit the multiple available paths in a NoC for best-effort and safety-critical senders, improving the average performance of BE. It couples the path selection process with temporal flow control based on the current state of the NoC. Chapters 4, 5, and 6 are based on this work.

[117]: A. Kostrzewa, S. Tobuschat, P. Axer, and R. Ernst. "Supervised sharing of virtual channels in Networks -on-Chip". In: *SIES*. 2014. DOI: `10.1109/SIES.2014.6871197`
This work presents the challenges and sketches solutions for global, dynamic and work-conserving arbitration in Networks-on-Chip using a control layer. Chapters 4 is partially based on this work.

[120]: A. Kostrzewa, S. Tobuschat, R. Ernst, and S. Saidi. "Safe and dynamic traffic rate control for networks-on-chips". In: *2016 Tenth IEEE/ACM International Symposium on Networks-on-Chip (NOCS)*. Aug. 2016, pages 1–8. DOI: `10.1109/NOCS.2016.7579321`
This work presents a global and dynamic admission control that controls the rates at which running applications can access the NoC. The mechanism allows enforcing behavioural models for different data streams as well as to dynamically adapt the rates values to the number of currently active applications. Chapters 5 is partially based on this work.

[121]: A. Kostrzewa, S. Tobuschat, L. Ecco, and R. Ernst. "Adaptive load distribution in mixed-critical Networks-on-Chip". In: *2017 22nd Asia and South Pacific Design Automation Conference (ASP-DAC)*. Jan. 2017, pages 732–737. DOI: `10.1109/ASPDAC.2017.7858411`
This work presents a protocol-based adaptive load distribution, which, by selectively de-touring BE traffic, allows to significantly improve the performance of the NoC. Chapters 4 is partially based on this work.

[124]: A. Kostrzewa, S. Tobuschat, and R. Ernst. "Self-Aware Network-On-Chip Control in Real-Time Systems". In: *IEEE Design & Test* 35.5 (Oct. 2018), pages 19–27. ISSN: 2168-2356. DOI: `10.1109/MDAT.2017.2763598`

This work discusses the application of the control layer in highly complex embedded applications, e.g., autonomous driving, which require simultaneously high performance and safety in highly dynamic setups. Chapters 4 is partially based on this work.

[162]: B. Nikolić, S. Tobuschat, L. Soares Indrusiak, R. Ernst, and A. Burns. "Real-time analysis of priority-preemptive NoCs with arbitrary buffer sizes and router delays". In: *Real-Time Systems* (June 2018). ISSN: 1573-1383. DOI: 10.1007/s11241-018-9312-0. URL: https://doi.org/10.1007/s11241-018-9312-0
This work presents an analysis method for computing the worst-case traversal times (WCTTs) of communication traffic flows accounting for backpressure. The approach is applicable to workloads deployed upon priority-preemptive NoCs using wormhole switching and per-traffic flow dedicated virtual channels. Additionally, the experimental evaluation led to an interesting finding that bigger virtual channel buffers do not necessarily yield better results, and in many cases can be counter-productive.

A.1.2 Unreviewed

[125]: A. Kostrzewa, S. Tobuschat, and R. Ernst. *Technical Report: Resource Manager and Control Layer, Basic Concepts*. Technical report. Institut für Datentechnik und Kommunikationsnetze, Feb. 2018
This technical report presents the control layer and its application for safety-critical real-time NoCs. Chapters 4 and 5 are based on this work.

A.2 Unrelated Publications

[70]: L. Ecco, S. Tobuschat, S. Saidi, and R. Ernst. "A mixed critical memory controller using bank privatization and fixed priority scheduling". In: *2014 IEEE 20th International Conference on Embedded and Real-Time Computing Systems and Applications*. Aug. 2014, pages 1–10. DOI: 10.1109/RTCSA.2014.6910550
This work presents a memory controller for mixed-critical workloads. The memory controller allows a predetermined number of critical and non-critical applications to coexist, while providing an interference-free memory for the former.

[218]: S. Tobuschat, R. Ernst, A. Hamann, and D. Ziegenbein. "System-level timing feasibility test for cyber-physical automotive systems". In: *2016*

11th IEEE Symposium on Industrial Embedded Systems (SIES). May 2016, pages 1–10. DOI: 10.1109/SIES.2016.7509419
This work presents a system-level timing feasibility test exploiting the robustness of control applications. It extends the worst-case and typical worst-case analysis (TWCA) to cope with to cope with periodic tasks that have varying execution times. Using the cumulative execution time of a task, instead of assuming the worst-case execution time for each activation, it improves the tightness of the worst-case response-times, to lessen the gap between analysis and extensive simulations. By taking robustness constraints into account for the system feasibility, it further decreases this gap.

[222]: S. Tobuschat, A. Kostrzewa, F. K. Bapp, and C. Dropmann. "Online monitoring for safety-critical multicore systems". In: *it - Information Technology* (2017). URL: https://doi.org/10.1515/itit-2017-0028
This work is an outcome of the ARAMiS research project. The work discusses the usage and limitation of monitoring approaches for safety-critical real-time system.

[195]: J. Schlatow, M. Mostl, S. Tobuschat, T. Ishigooka, and R. Ernst. "Data-Age Analysis and Optimisation for Cause-Effect Chains in Automotive Control Systems". In: *2018 IEEE 13th International Symposium on Industrial Embedded Systems (SIES)*. June 2018, pages 1–9. DOI: 10.1109/SIES.2018.8442077
This work addresses the latency analysis for multi-rate distributed cause-effect chains considering static-priority preemptive scheduling of offset-synchronised periodic tasks. The main contribution is a Mixed Integer Linear Program-based optimisation to select design parameters (priorities, task-to-processor mapping, offsets) that minimise the data age.

Bibliography

References

[1]   URL: https://semiengineering.com/the-race-to-autonomous-cars/ (cited on pages 3, 4).

[2]   *DO-254: Design Assurance Guidance for Airborne Electronic Hardware.* Apr. 2000 (cited on page 6).

[3]   *ISO 15005: Road vehicles - Ergonomic aspects of transport information and control systems - Dialogue management principles and compliance procedures.* 2002 (cited on page 7).

[4]   *ISO 16951: Road vehicles - Ergonomic aspects of transport information and control systems (TICS) - Procedures for determining priority of on-board messages presented to drivers.* 2004 (cited on page 7).

[5]   *DO-297: Integrated Modular Avionics (IMA) Development Guidance and Certifcation Considerations.* 2005 (cited on page 6).

[6]   *DO-178B: software considerations in airborne systems and equipment certification.* Dec. 2009 (cited on page 6).

[7]   *ISO 26262:2011, Road vehicles - Functional safety (2011).* 2011 (cited on pages 12, 46, 80, 130).

[8]   *AUTOSAR: Glossary.* AUTOSAR Consortium, May 2012. URL: www.autosar.org (cited on pages 238, 241).

[9]   *JESD79-3F: DDR3 SDRAM Specification.* JEDEC. Arlington, Va, USA: JEDEC Solid State Technology Association, July 2012 (cited on page 176).

[10] *AUTOSAR: E2E Protocol Specification.* AUTOSAR Consortium, Dec. 2017. URL: www.autosar.org (cited on pages 11, 12).

[11] "IEEE Standard for Local and Metropolitan Area Network–Bridges and Bridged Networks". In: *IEEE Std 802.1Q-2018 (Revision of IEEE Std 802.1Q-2014)* (July 2018), pages 1–1993. DOI: 10.1109/IEEESTD.2018.8403927 (cited on pages 239, 240).

[12] *OMNEST Website*. accessed on 28-05-2018. URL: https://www.omnest.com (cited on pages 42, 167).

[13] *ARP4754: Certifcation Considerations for Highly-Integrated Or Complex Aircraft Systems*. SAE International, 1996 (cited on page 6).

[14] *ARP4761: Guidelines and Methods for Conducting the Safety Assessment Process on Civil Airborne Systems and Equipment*. SAE International, 1996 (cited on page 6).

[15] *IEC 61508: Functional Safety of Electrical/Electronic/Programmable Electronic Safety Related Systems*. 1999 (cited on pages 4–6, 80, 130, 240).

[16] L. Abdallah, M. Jan, J. Ermont, and C. Fraboul. "I/O contention aware mapping of multi-criticalities real-time applications over many-core architectures". In: *22nd IEEE Real-Time and embedded Technology and Applications symposium (RTAS 2016)*. Vienna, Austria, Apr. 2016, pp. 25-28. URL: https://hal.archives-ouvertes.fr/hal-01445139 (cited on page 169).

[17] Adapteva. *Epiphany Architecture Reference*. Adapteva Inc. Mar. 2014. URL: http://www.adapteva.com/docs/epiphany_arch_ref.pdf (cited on pages 49, 54).

[18] A. Adriahantenaina, H. Charlery, A. Greiner, L. Mortiez, and C. A. Zeferino. "SPIN: a scalable, packet switched, on-chip micro-network". In: *2003 Design, Automation and Test in Europe Conference and Exhibition*. 2003, 70–73 suppl. DOI: 10.1109/DATE.2003.1253808 (cited on page 35).

[19] N. Agarwal, T. Krishna, L. S. Peh, and N. K. Jha. "GARNET: A detailed on-chip network model inside a full-system simulator". In: *2009 IEEE International Symposium on Performance Analysis of Systems and Software*. Apr. 2009, pages 33–42. DOI: 10.1109/ISPASS.2009.4919636 (cited on page 42).

[20] M. Alho and J. Nurmi. "Implementation of Interface Router Ip for Proteo Network-on-chip". In: 2003 (cited on page 34).

[21] K. Al-Tawil, M. Abd-El-Barr, and F. Ashraf. "A survey and comparison of wormhole routing techniques in a mesh networks". In: *Network, IEEE* 11.2 (Apr. 1997), pages 38–45. ISSN: 0890-8044. DOI: 10.1109/65.580917 (cited on page 25).

[22] A. Argarwal, C. Iskander, and R. Shankar. "Survey of Network on Chip (NoC) Architectures & Contributions". In: (2009) (cited on page 19).

[23] Arteris. *A comparison of network-on-chip and busses (whitepaper)*. 2005. URL: www.arteris.com (cited on page 31).

[24] P. Axer, J. Diemer, M. Negrean, M. Sebastian, S. Schliecker, and R. Ernst. "Mastering MPSoCs for Mixed-critical Applications". In: *IPSJ Transactions on System LSI Design Methodology* 4 (2011), pages 91–116. DOI: 10.2197/ ipsjtsldm.4.91 (cited on pages 8, 13).

[25] J. Bainbridge and S. Furber. "Chain: a delay-insensitive chip area interconnect". In: *IEEE Micro* 22.5 (Sept. 2002), pages 16–23. ISSN: 0272-1732. DOI: 10.1109/MM.2002.1044296 (cited on page 31).

[26] J. Balfour and W. J. Dally. "Design Tradeoffs for Tiled CMP On-chip Networks". In: *Proceedings of the 20th Annual International Conference on Supercomputing*. ICS '06. Cairns, Queensland, Australia: ACM, 2006, pages 187–198. ISBN: 1-59593-282-8. DOI: 10.1145/1183401.1183430. URL: http://doi.acm.org/10.1145/1183401.1183430 (cited on page 142).

[27] S. Balakrishnan and F. Ozguner. "A priority-driven flow control mechanism for real-time traffic in multiprocessor networks". In: *IEEE Transactions on Parallel and Distributed Systems* 9.7 (July 1998), pages 664–678. ISSN: 1045-9219. DOI: 10.1109/71.707545 (cited on page 48).

[28] S. Baruah, H. Li, and L. Stougie. "Towards the Design of Certifiable Mixed-criticality Systems". In: *Real-Time and Embedded Technology and Applications Symposium (RTAS), 2010 16th IEEE*. 2010, pages 13–22. DOI: 10.1109/RTAS.2010.10 (cited on pages 1, 79).

[29] S. Baruah, A. Burns, and R. Davis. "Response-Time Analysis for Mixed Criticality Systems". In: *Real-Time Systems Symposium (RTSS), 2011 IEEE 32nd*. Nov. 2011, pages 34–43. DOI: 10.1109/RTSS.2011.12 (cited on page 82).

[30] D. U. Becker. "Efficient microarchitecture for network-on-chip routers". In: *PhD dissertation, Stanford University, Dept. of Electrical Engineering* (2012), pages 11–27. URL: http://purl.stanford.edu/wr368td5072 (cited on pages 89, 90).

[31] L. Benini and G. D. Micheli. "Networks on chips: a new SoC paradigm". In: *Computer* 35.1 (Jan. 2002), pages 70–78. ISSN: 0018-9162. DOI: 10.1109/ 2.976921 (cited on page 5).

[32] Y. Ben-Itzhak, E. Zahavi, I. Cidon, and A. Kolodny. "HNOCS: Modular open-source simulator for Heterogeneous NoCs". In: *Embedded Computer Systems (SAMOS), 2012 International Conference on*. July 2012, pages 51–57. DOI: 10.1109/SAMOS.2012.6404157 (cited on pages 42, 43, 72, 96, 112, 167).

[33] D. Bertozzi and L. Benini. "Xpipes: a network-on-chip architecture for gigas-cale systems-on-chip". In: *IEEE Circuits and Systems Magazine* 4.2 (2004), pages 18–31. ISSN: 1531-636X. DOI: 10.1109/MCAS.2004.1330747 (cited on page 37).

[34] D. Bertozzi, L. Benini, and G. D. Micheli. "Error control schemes for on-chip communication links: the energy-reliability tradeoff". In: *IEEE Transactions on Computer-Aided Design of Integrated Circuits and Systems* 24.6 (June 2005), pages 818–831. ISSN: 0278-0070. DOI: 10.1109/TCAD.2005.847907 (cited on page 148).

[35] T. Bjerregaard and J. Sparsoe. "Scheduling discipline for latency and band-width guarantees in asynchronous network-on-chip". In: *Asynchronous Circuits and Systems, 2005. ASYNC 2005. Proceedings. 11th IEEE International Symposium on*. Mar. 2005, pages 34–43. DOI: 10.1109/ASYNC.2005.7 (cited on page 82).

[36] T. Bjerregaard and S. Mahadevan. "A Survey of Research and Practices of Network-on-chip". In: *ACM Comput. Surv.* 38.1 (June 2006). ISSN: 0360-0300. DOI: 10.1145/1132952.1132953 (cited on page 19).

[37] T. Bjerregaard and J. Sparso. "Implementation of guaranteed services in the MANGO clockless network-on-chip". In: *Computers and Digital Techniques, IEE Proceedings-*. Volume 153. 4. IET. 2006, pages 217–229 (cited on pages 25, 33).

[38] E. Bolotin, I. Cidon, R. Ginosar, and A. Kolodny. "Cost considerations in network on chip". In: *Integration, the VLSI Journal* 38.1 (2004), pages 19–42. ISSN: 0167-9260. DOI: https://doi.org/10.1016/j.vlsi.2004.03.006. URL: http://www.sciencedirect.com/science/article/pii/S0167926004000343 (cited on page 34).

[39] E. Bolotin, I. Cidon, R. Ginosar, and A. Kolodny. "QNoC: QoS Architecture and Design Process for Network on Chip". In: *J. Syst. Archit.* 50.2-3 (Feb. 2004), pages 105–128. ISSN: 1383-7621. DOI: 10.1016/j.sysarc.2003.07.004 (cited on pages 34, 82, 95, 111).

[40] S. Borkar. "Thousand Core Chips: A Technology Perspective". In: *Proceedings of the 44th Annual Design Automation Conference*. DAC '07. San Diego, California: ACM, 2007, pages 746–749. ISBN: 978-1-59593-627-1. DOI: 10.1145/1278480.1278667. URL: http://doi.acm.org/10.1145/1278480.1278667 (cited on page 140).

[41] S. Borkar. "Future of Interconnect Fabric: A Contrarian View". In: *Proceedings of the 12th ACM/IEEE International Workshop on System Level Interconnect Prediction*. SLIP '10. Anaheim, California, USA: ACM, 2010, pages 1–2. ISBN: 978-1-4503-0037-7. DOI: 10.1145/1811100.1811101.

URL: http://doi.acm.org/10.1145/1811100.1811101 (cited on page 140).

[42]   M. Boyer, B. Dupont De Dinechin, A. Graillat, and L. Havet. "Computing Routes and Delay Bounds for the Network-on-Chip of the Kalray MPPA2 Processor". In: *ERTS 2018 - 9th European Congress on Embedded Real Time Software and Systems*. Toulouse, France, Jan. 2018. URL: https://hal.archives-ouvertes.fr/hal-01707911 (cited on pages 33, 78, 122, 142).

[43]   G. Buja, A. Zuccollo, and J. Pimentel. "Overcoming babbling-idiot failures in the FlexCAN architecture: a simple bus-guardian". In: *2005 IEEE Conference on Emerging Technologies and Factory Automation*. Volume 2. Sept. 2005, 8 pp.-468. DOI: 10.1109/ETFA.2005.1612713 (cited on page 237).

[44]   A. Burns, J. Harbin, and L. Indrusiak. "A Wormhole NoC Protocol for Mixed Criticality Systems". In: *Real-Time Systems Symposium (RTSS), 2014 IEEE*. Dec. 2014, pages 184–195. DOI: 10.1109/RTSS.2014.13 (cited on pages 82, 83, 172).

[45]   A. Burns and R. Davis. "Mixed criticality systems-a review (7-th Ed)". In: *Department of Computer Science, University of York, Tech. Rep* (Jan. 2016). URL: https://www-users.cs.york.ac.uk/burns/review.pdf (cited on page 79).

[46]   E. Carara, F. Moraes, and N. Calazans. "Router Architecture for High-performance NoCs". In: *Proceedings of the 20th Annual Conference on Integrated Circuits and Systems Design*. SBCCI '07. Copacabana, Rio de Janeiro: ACM, 2007, pages 111–116. ISBN: 978-1-59593-816-9. DOI: 10.1145/1284480.1284515. URL: http://doi.acm.org/10.1145/1284480.1284515 (cited on page 142).

[47]   A. do Carmo Lucas, S. Heithecker, and R. Ernst. "FlexWAFE - A High-end Real-Time Stream Processing Library for FPGAs". In: *2007 44th ACM/IEEE Design Automation Conference*. June 2007, pages 916–921. DOI: 10.1109/DAC.2007.375295 (cited on pages 176, 177).

[48]   V. Catania, A. Mineo, S. Monteleone, M. Palesi, and D. Patti. "Cycle-Accurate Network on Chip Simulation with Noxim". In: *ACM Trans. Model. Comput. Simul.* 27.1 (Aug. 2016), 4:1–4:25. ISSN: 1049-3301. DOI: 10.1145/2953878. URL: http://doi.acm.org/10.1145/2953878 (cited on page 42).

[49]   S. Chakraborty, S. Kunzli, and L. Thiele. "A general framework for analysing system properties in platform-based embedded system designs". In: *Design, Automation and Test in Europe Conference and Exhibition, 2003*. 2003, pages 190–195. DOI: 10.1109/DATE.2003.1253607 (cited on page 50).

[50] P. Christensson. *SoC Definition*. Nov. 2015. URL: https://techterms.
      com/definition/soc (cited on page 240).

[51] K. Chrysanthou, P. Englezakis, A. Prodromou, A. Panteli, C. Nicopoulos,
      Y. Sazeides, and G. Dimitrakopoulos. "An Online and Real-Time Fault
      Detection and Localization Mechanism for Network-on-Chip Architectures".
      In: *ACM Trans. Archit. Code Optim.* 13.2 (June 2016), 22:1–22:26. ISSN:
      1544-3566. DOI: 10.1145/2930670. URL: http://doi.acm.org/10.
      1145/2930670 (cited on page 201).

[52] Commission, International Electrotechnical. *Electropedia: The World's On-
      line Electrotechnical Vocabulary*. 2008. URL: http://www.electropedia.
      org/ (cited on pages 237–240).

[53] R. L. Cruz. "A calculus for network delay. II. Network analysis". In: *IEEE
      Transactions on Information Theory* 37.1 (Jan. 1991), pages 132–141. ISSN:
      0018-9448. DOI: 10.1109/18.61110 (cited on page 41).

[54] W. Dally and B. Towles. *Principles and Practices of Interconnection Net-
      works*. San Francisco, CA, USA: Morgan Kaufmann Publishers Inc., 2003.
      ISBN: 0122007514 (cited on pages 13, 22, 27, 37, 85, 119, 121, 238, 240).

[55] M. Dall'Osso, G. Biccari, L. Giovannini, D. Bertozzi, and L. Benini. "Xpipes:
      A latency insensitive parameterized network-on-chip architecture for multi-
      processor SoCs". In: *2012 IEEE 30th International Conference on Computer
      Design (ICCD)*. Sept. 2012, pages 45–48. DOI: 10.1109/ICCD.2012.
      6378615 (cited on page 37).

[56] D. Dasari, B. Nikoli'c, V. N'elis, and S. M. Petters. "NoC Contention
      Analysis Using a Branch-and-prune Algorithm". In: *ACM Trans. Embed.
      Comput. Syst.* 13.3s (Mar. 2014), 113:1–113:26. ISSN: 1539-9087. DOI:
      10.1145/2567937. URL: http://doi.acm.org/10.1145/2567937
      (cited on page 48).

[57] M. Dehbashi and G. Fey. *Debug Automation from Pre-Silicon to Post-Silicon*.
      Jan. 2015, pages 1–171 (cited on page 202).

[58] J. Diemer and R. Ernst. "Back Suction: Service Guarantees for Latency-
      Sensitive On-chip Networks". In: *Networks-on-Chip (NOCS), 2010 Fourth
      ACM/IEEE International Symposium on*. 2010, pages 155–162. DOI: 10.
      1109/NOCS.2010.38 (cited on pages 32, 83, 105, 111).

[59] J. Diemer, R. Ernst, and M. Kauschke. "Efficient throughput-guarantees
      for latency-sensitive networks-on-chip". In: *Design Automation Conference
      (ASP-DAC), 2010 15th Asia and South Pacific*. 2010, pages 529–534. DOI:
      10.1109/ASPDAC.2010.5419828 (cited on pages 83, 238).

[60]   J. Diemer, J. Rox, M. Negrean, S. Stein, and R. Ernst. "Real-time Communi-
       cation Analysis for Networks with Two-stage Arbitration". In: *Proceedings
       of the Ninth ACM International Conference on Embedded Software*. EM-
       SOFT '11. Taipei, Taiwan: ACM, 2011, pages 243–252. ISBN: 978-1-4503-
       0714-7. DOI: 10.1145/2038642.2038680 (cited on page 52).

[61]   J. Diemer, D. Thiele, and R. Ernst. "Formal worst-case timing analysis of
       Ethernet topologies with strict-priority and AVB switching". In: *Industrial
       Embedded Systems (SIES), 2012 7th IEEE International Symposium on*. June
       2012, pages 1–10. DOI: 10.1109/SIES.2012.6356564 (cited on pages 68,
       69).

[62]   J. Diemer, P. Axer, and R. Ernst. "Compositional Performance Analysis
       in Python with pyCPA". In: *3rd International Workshop on Analysis Tools
       and Methodologies for Embedded and Real-time Systems (WATERS)*. July
       2012. URL: http://retis.sssup.it/waters2012/WATERS-2012-
       Proceedings.pdf (cited on page 178).

[63]   J. F. Diemer. "Predictable Architecture and Performance Analysis for
       General-Purpose Networks-on-Chip". PhD thesis. TU Braunschweig, 2016
       (cited on pages 52, 54, 70, 71, 238, 240).

[64]   G. Dimitrakopoulos, N. Chrysos, and K. Galanopoulos. "Fast arbiters for
       on-chip network switches". In: *2008 IEEE International Conference on
       Computer Design*. Oct. 2008, pages 664–670. DOI: 10.1109/ICCD.2008.
       4751932 (cited on pages 89, 90).

[65]   B. de Dinechin, R. Ayrignac, P.-E. Beaucamps, P. Couvert, B. Ganne, P. de
       Massas, F. Jacquet, S. Jones, N. Chaisemartin, F. Riss, and T. Strudel. "A
       clustered manycore processor architecture for embedded and accelerated ap-
       plications". In: *High Performance Extreme Computing Conference (HPEC),
       2013 IEEE*. Sept. 2013, pages 1–6. DOI: 10.1109/HPEC.2013.6670342
       (cited on pages 33, 78, 121, 122, 142).

[66]   B. D. de Dinechin and A. Graillat. "Feed-Forward Routing for the Worm-
       hole Switching Network-on-Chip of the Kalray MPPA2 Processor". In:
       *Proceedings of the 10th International Workshop on Network on Chip Ar-
       chitectures*. NoCArc'17. Cambridge, MA, USA: ACM, 2017, 10:1–10:6.
       ISBN: 978-1-4503-5542-1. DOI: 10.1145/3139540.3139542. URL: http:
       //doi.acm.org/10.1145/3139540.3139542 (cited on page 33).

[67]   B. D. de Dinechin and A. Graillat. "Network-on-chip Service Guarantees on
       the Kalray MPPA-256 Bostan Processor". In: *Proceedings of the 2Nd Inter-
       national Workshop on Advanced Interconnect Solutions and Technologies for
       Emerging Computing Systems*. AISTECS '17. Stockholm, Sweden: ACM,
       2017, pages 35–40. ISBN: 978-1-4503-5226-0. DOI: 10.1145/3073763.

3073770. URL: http://doi.acm.org/10.1145/3073763.3073770 (cited on page 33).

[68]   R. Dobkin, R. Ginosar, and I. Cidon. "QNoC Asynchronous Router with Dynamic Virtual Channel Allocation". In: *First International Symposium on Networks-on-Chip (NOCS'07)*. May 2007, pages 218–218. DOI: 10.1109/ NOCS.2007.36 (cited on page 34).

[69]   J. Duato, S. Yalamanchili, and L. Ni. *Interconnection Networks*. San Francisco, CA, USA: Morgan Kaufmann Publishers Inc., 2003. ISBN: 9780080508993 (cited on pages 13, 22, 37, 119, 121, 144, 238, 240).

[70]   L. Ecco, S. Tobuschat, S. Saidi, and R. Ernst. "A mixed critical memory controller using bank privatization and fixed priority scheduling". In: *2014 IEEE 20th International Conference on Embedded and Real-Time Computing Systems and Applications*. Aug. 2014, pages 1–10. DOI: 10.1109/ RTCSA.2014.6910550 (cited on page 206).

[71]   G. A. Elliott and J. H. Anderson. "Real-World Constraints of GPUs in Real-Time Systems". In: *2011 IEEE 17th International Conference on Embedded and Real-Time Computing Systems and Applications*. Volume 2. Aug. 2011, pages 48–54. DOI: 10.1109/RTCSA.2011.46 (cited on page 9).

[72]   M. Fattah, M. Daneshtalab, P. Liljeberg, and J. Plosila. "Smart hill climbing for agile dynamic mapping in many-core systems". In: *2013 50th ACM/EDAC/IEEE Design Automation Conference (DAC)*. May 2013, pages 1–6. DOI: 10.1145/2463209.2488782 (cited on page 169).

[73]   T. Felicijan, J. Bainbridge, and S. Furber. "An asynchronous low latency arbiter for Quality of Service (QoS) applications". In: *Proceedings of the 12th IEEE International Conference on Fuzzy Systems (Cat. No.03CH37442)*. Dec. 2003, pages 123–126. DOI: 10.1109/ICM.2003.1287737 (cited on page 31).

[74]   F. Feliciian and S. B. Furber. "An asynchronous on-chip network router with quality-of-service (QoS) support". In: *IEEE International SOC Conference, 2004. Proceedings*. Sept. 2004, pages 274–277. DOI: 10.1109/SOCC.2004. 1362432 (cited on page 31).

[75]   T. Ferrandiz, F. Frances, and C. Fraboul. "A method of computation for worst-case delay analysis on SpaceWire networks". In: *2009 IEEE International Symposium on Industrial Embedded Systems*. July 2009, pages 19–27. DOI: 10.1109/SIES.2009.5196187 (cited on page 48).

[76]   T. Ferrandiz, F. Frances, and C. Fraboul. "A Sensitivity Analysis of Two Worst-Case Delay Computation Methods for SpaceWire Networks". In: *2012 24th Euromicro Conference on Real-Time Systems*. July 2012, pages 47–56. DOI: 10.1109/ECRTS.2012.35 (cited on page 48).

[77]   T. Forest and M. Jochim. "On the Fault Detection Capabilities of AU-
       TOSAR's End-to-End Communication Protection CRC's". In: *SAE 2011
       World Congress & Exhibition*. SAE International, Apr. 2011. DOI: https:
       //doi.org/10.4271/2011-01-0999. URL: https://doi.org/10.
       4271/2011-01-0999 (cited on page 11).

[78]   S. Friederich, J. Heisswolf, and J. Becker. "Hardware/software debugging of
       large scale many-core architectures". In: *2014 27th Symposium on Integrated
       Circuits and Systems Design (SBCCI)*. Sept. 2014, pages 1–7. DOI: 10.
       1145/2660540.2661013 (cited on page 202).

[79]   S. Friederich, M. Neber, and J. Becker. "Power Management Controller for
       Online Power Saving in Network-on-Chips". In: *2016 IEEE 10th Interna-
       tional Symposium on Embedded Multicore/Many-core Systems-on-Chip (MC-
       SOC)*. Volume 00. Sept. 2016, pages 109–116. DOI: 10.1109/MCSoC.2016.
       22. URL: doi.ieeecomputersociety.org/10.1109/MCSoC.2016.22
       (cited on page 200).

[80]   D. Göhringer, L. Meder, O. Oey, and J. Becker. "Reliable and Adaptive
       Network-on-chip Architectures for Cyber Physical Systems". In: *ACM Trans.
       Embed. Comput. Syst.* 12.1s (Mar. 2013), 51:1–51:21. ISSN: 1539-9087. DOI:
       10.1145/2435227.2435247. URL: http://doi.acm.org/10.1145/
       2435227.2435247 (cited on page 148).

[81]   K. Goossens, J. Dielissen, and A. Radulescu. "Æthereal network on chip:
       concepts, architectures, and implementations". In: *Design & Test of Comput-
       ers, IEEE* 22.5 (2005), pages 414–421 (cited on pages 25, 31, 122).

[82]   K. Goossens, B. Vermeulen, R. v. Steeden, and M. Bennebroek. "Transaction-
       Based Communication-Centric Debug". In: *First International Symposium
       on Networks-on-Chip (NOCS'07)*. May 2007, pages 95–106. DOI: 10.1109/
       NOCS.2007.46 (cited on page 202).

[83]   K. Goossens, B. Vermeulen, and A. B. Nejad. "A high-level debug envi-
       ronment for communication-centric debug". In: *2009 Design, Automation
       Test in Europe Conference Exhibition*. Apr. 2009, pages 202–207. DOI:
       10.1109/DATE.2009.5090658 (cited on page 202).

[84]   K. Goossens and A. Hansson. "The Aethereal Network on Chip After
       Ten Years: Goals, Evolution, Lessons, and Future". In: *Proceedings of the
       47th Design Automation Conference*. DAC '10. Anaheim, California: ACM,
       2010, pages 306–311. ISBN: 978-1-4503-0002-5. DOI: 10.1145/1837274.
       1837353 (cited on page 81).

[85]   P. Gratz, C. Kim, R. McDonald, S. W. Keckler, and D. Burger. "Imple-
       mentation and Evaluation of On-Chip Network Architectures". In: *2006
       International Conference on Computer Design*. Oct. 2006, pages 477–484.
       DOI: 10.1109/ICCD.2006.4380859 (cited on page 140).

[86]   D. Gross, J. F. Shortle, J. M. Thompson, and C. M. Harris. *Fundamentals of Queueing Theory*. 4th. New York, NY, USA: Wiley-Interscience, 2008. ISBN: 047179127X, 9780471791270 (cited on page 41).

[87]   B. Grot, J. Hestness, S. Keckler, and O. Mutlu. "Kilo-NOC: A heterogeneous network-on-chip architecture for scalability and service guarantees". In: *Computer Architecture (ISCA), 2011 38th Annual International Symposium on*. June 2011, pages 401–412 (cited on pages 82, 95).

[88]   P. Guerrier and A. Greiner. "A generic architecture for on-chip packet-switched interconnections". In: *Design, Automation and Test in Europe Conference and Exhibition 2000. Proceedings*. 2000, pages 250–256. DOI: 10.1109/DATE.2000.840047 (cited on page 35).

[89]   A. Hansson, M. Coenen, and K. Goossens. "Channel trees: Reducing latency by sharing time slots in time-multiplexed Networks on Chip". In: *CODES+ISSS*. Sept. 2007, pages 149–154 (cited on page 81).

[90]   A. Hansson, M. Subburaman, and K. Goossens. "Aelite: A flit-synchronous Network on Chip with composable and predictable services". In: *2009 Design, Automation Test in Europe Conference Exhibition*. Apr. 2009, pages 250–255. DOI: 10.1109/DATE.2009.5090666 (cited on page 31).

[91]   Y. Hara, H. Tomiyama, S. Honda, and H. Takada. "Proposal and Quantitative Analysis of the CHStone Benchmark Program Suite for Practical C-based High-level Synthesis". In: *Journal of Information Processing*. 17 (2009), pages 242–254 (cited on pages 98, 114).

[92]   M. Harrand and Y. Durand. "Network on chip with quality of service". US Patent 8,619,622. Dec. 2013 (cited on pages 49, 54).

[93]   J. Heißwolf, R. König, and J. Becker. "A Scalable NoC Router Design Providing QoS Support Using Weighted Round Robin Scheduling". In: *2012 IEEE 10th International Symposium on Parallel and Distributed Processing with Applications*. July 2012, pages 625–632. DOI: 10.1109/ISPA.2012.93 (cited on pages 32, 33, 82).

[94]   J. Heißwolf, A. Zaib, A. Weichslgartner, R. König, T. Wild, J. Teich, A. Herkersdorf, and J. Becker. "Hardware-assisted Decentralized Resource Management for Networks on Chip with QoS". In: *2012 IEEE 26th International Parallel and Distributed Processing Symposium Workshops PhD Forum*. May 2012, pages 234–241. DOI: 10.1109/IPDPSW.2012.25 (cited on page 123).

[95]   J. Heisswolf, A. Zaib, A. Weichslgartner, R. König, T. Wild, J. Teich, A. Herkersdorf, and J. Becker. "Virtual Networks – Distributed Communication Resource Management". In: *ACM Trans. Reconfigurable Technol. Syst.* 6.2 (Aug. 2013), 8:1–8:14. ISSN: 1936-7406. DOI: 10.1145/2492186. URL: http://doi.acm.org/10.1145/2492186 (cited on pages 123, 145, 199).

[96]   J. Heisswolf, A. Zaib, A. Weichslgartner, M. Karle, M. Singh, T. Wild, J. Teich, A. Herkersdorf, and J. Becker. "The Invasive Network on Chip - A Multi-Objective Many-Core Communication Infrastructure". In: *ARCS 2014; 2014 Workshop Proceedings on Architecture of Computing Systems.* Feb. 2014, pages 1–8 (cited on page 32).

[97]   J. Heißwolf. "A Scalable and Adaptive Network on Chip for Many-Core Architectures". PhD thesis. 2014 (cited on page 32).

[98]   J. Heisswolf, A. Weichslgartner, A. Zaib, S. Friederich, L. Masing, C. Stein, M. Duden, R. Klöpfer, J. Teich, T. Wild, A. Herkersdorf, and J. Becker. "Fault-tolerant communication in invasive networks on chip". In: *2015 NASA/ESA Conference on Adaptive Hardware and Systems (AHS).* June 2015, pages 1–8. DOI: 10.1109/AHS.2015.7231156 (cited on page 201).

[99]   R. Henia, A. Hamann, M. Jersak, R. Racu, K. Richter, and R. Ernst. "System level performance analysis - the SymTA/S approach". In: *Computers and Digital Techniques, IEE Proceedings* - 152.2 (Mar. 2005), pages 148–166. ISSN: 1350-2387. DOI: 10.1049/ip-cdt:20045088 (cited on pages 7, 41, 49–51, 54, 71, 122).

[100]   R. Hofmann, L. Ahrendts, and R. Ernst. "CPA: Compositional Performance Analysis". In: *Handbook of Hardware/Software Codesign.* Edited by S. Ha and J. Teich. Dordrecht: Springer Netherlands, 2017, pages 721–751. ISBN: 978-94-017-7267-9. DOI: 10.1007/978-94-017-7267-9_24. URL: https://doi.org/10.1007/978-94-017-7267-9_24 (cited on page 49).

[101]   Y. Hoskote, S. Vangal, A. Singh, N. Borkar, and S. Borkar. "A 5-GHz Mesh Interconnect for a Teraflops Processor". In: *IEEE Micro* 27.5 (Sept. 2007), pages 51–61. ISSN: 0272-1732. DOI: 10.1109/MM.2007.4378783 (cited on page 140).

[102]   J. Hu and R. Marculescu. "DyAD: Smart Routing for Networks-on-chip". In: *Proceedings of the 41st Annual Design Automation Conference.* DAC '04. San Diego, CA, USA: ACM, 2004, pages 260–263. ISBN: 1-58113-828-8. DOI: 10.1145/996566.996638 (cited on page 32).

[103] L. Indrusiak, J. Harbin, and A. Burns. "Average and Worst-Case Latency Improvements in Mixed-Criticality Wormhole Networks-on-Chip". In: *Real-Time Systems (ECRTS), 2015 27th Euromicro Conference on*. July 2015, pages 47–56. DOI: 10.1109/ECRTS.2015.12 (cited on pages 82, 83, 86, 95, 172).

[104] L. Jain, B. Al-Hashimi, M. Gaur, V. Laxmi, and A. Narayanan. "NIRGAM: a simulator for NoC interconnect routing and application modeling". In: *Workshop on Diagnostic Services in Network-on-Chips, Design, Automation and Test in Europe Conference (DATE' 07)*. Apr. 2007. URL: http:// nirgam.ecs.soton.ac.uk (cited on page 42).

[105] N. Jiang, J. Balfour, D. U. Becker, B. Towles, W. J. Dally, G. Michelogiannakis, and J. Kim. "A detailed and flexible cycle-accurate Network-on-Chip simulator". In: *2013 IEEE International Symposium on Performance Analysis of Systems and Software (ISPASS)*. Apr. 2013, pages 86–96. DOI: 10.1109/ISPASS.2013.6557149 (cited on page 42).

[106] M. R. John, R. James, J. Jose, E. Isaac, and J. K. Antony. "A Novel Energy Efficient Source Routing for Mesh NoCs". In: *2014 Fourth International Conference on Advances in Computing and Communications*. Aug. 2014, pages 125–129. DOI: 10.1109/ICACC.2014.36 (cited on page 22).

[107] A. B. Kahng, B. Li, L. Peh, and K. Samadi. "ORION 2.0: A Power-Area Simulator for Interconnection Networks". In: *IEEE Transactions on Very Large Scale Integration (VLSI) Systems* 20.1 (Jan. 2012), pages 191–196. ISSN: 1063-8210. DOI: 10.1109/TVLSI.2010.2091686 (cited on page 42).

[108] H. M. Kamali and S. Hessabi. "AdapNoC: A fast and flexible FPGA-based NoC simulator". In: *2016 26th International Conference on Field Programmable Logic and Applications (FPL)*. Aug. 2016, pages 1–8. DOI: 10.1109/FPL.2016.7577377 (cited on page 42).

[109] K. Kariniemi and J. Nurmi. "Fault tolerant XGFT network on chip for multi processor system on chip circuits". In: *International Conference on Field Programmable Logic and Applications, 2005*. Aug. 2005, pages 203–210. DOI: 10.1109/FPL.2005.1515723 (cited on page 37).

[110] H. Kashif, S. Gholamian, and H. Patel. "SLA: A Stage-Level Latency Analysisfor Real-Time Communicationin a Pipelined Resource Model". In: *IEEE Transactions on Computers* 64.4 (Apr. 2015), pages 1177–1190. ISSN: 0018-9340. DOI: 10.1109/TC.2014.2315617 (cited on pages 48, 49).

[111] H. Kashif and H. Patel. "Buffer Space Allocation for Real-Time Priority-Aware Networks". In: *2016 IEEE Real-Time and Embedded Technology and Applications Symposium (RTAS)*. Apr. 2016, pages 1–12. DOI: 10.1109/ RTAS.2016.7461324 (cited on pages 49, 105).

[112]   J. O. Kephart and D. M. Chess. "The vision of autonomic computing". In: *Computer* 36.1 (Jan. 2003), pages 41–50. ISSN: 0018-9162. DOI: 10.1109/ MC.2003.1160055 (cited on page 199).

[113]   B. Kienhuis, E. Deprettere, K. Vissers, and P. V. D. Wolf. "An approach for quantitative analysis of application-specific dataflow architectures". In: *Proceedings IEEE International Conference on Application-Specific Systems, Architectures and Processors*. July 1997, pages 338–349. DOI: 10.1109/ ASAP.1997.606839 (cited on pages 46, 47).

[114]   J. Kim and H. Shin. *Algorithm & SoC Design for Automotive Vision Systems: For Smart Safe Driving System*. Springer Publishing Company, Incorporated, 2014. ISBN: 9401790744, 9789401790741 (cited on page 9).

[115]   B. Kim, J. Kim, S. Hong, and S. Lee. "A real-time communication method for wormhole switching networks". In: *Parallel Processing, 1998. Proceedings. 1998 International Conference on*. Aug. 1998, pages 527–534. DOI: 10. 1109/ICPP.1998.708526 (cited on page 48).

[116]   P. Koopman and T. Chakravarty. "Cyclic redundancy code (CRC) polynomial selection for embedded networks". In: *International Conference on Dependable Systems and Networks, 2004*. June 2004, pages 145–154. DOI: 10.1109/DSN.2004.1311885 (cited on pages 148, 154).

[117]   A. Kostrzewa, S. Tobuschat, P. Axer, and R. Ernst. "Supervised sharing of virtual channels in Networks -on-Chip". In: *SIES*. 2014. DOI: 10.1109/ SIES.2014.6871197 (cited on page 205).

[118]   A. Kostrzewa, S. Saidi, L. Ecco, and R. Ernst. "Dynamic admission control for real-time networks-on-chips". In: *2016 21ˢᵗ Asia and South Pacific Design Automation Conference (ASP-DAC)*. Jan. 2016, pages 719–724. DOI: 10.1109/ASPDAC.2016.7428096 (cited on page 86).

[119]   A. Kostrzewa, S. Saidi, and R. Ernst. "Slack-based resource arbitration for real-time Networks-on-Chip". In: *2016 Design, Automation Test in Europe Conference Exhibition (DATE)*. Mar. 2016, pages 1012–1017 (cited on pages 168, 172).

[120]   A. Kostrzewa, S. Tobuschat, R. Ernst, and S. Saidi. "Safe and dynamic traffic rate control for networks-on-chips". In: *2016 Tenth IEEE/ACM International Symposium on Networks-on-Chip (NOCS)*. Aug. 2016, pages 1–8. DOI: 10.1109/NOCS.2016.7579321 (cited on pages 136, 205).

[121]   A. Kostrzewa, S. Tobuschat, L. Ecco, and R. Ernst. "Adaptive load distribution in mixed-critical Networks-on-Chip". In: *2017 22nd Asia and South Pacific Design Automation Conference (ASP-DAC)*. Jan. 2017, pages 732– 737. DOI: 10.1109/ASPDAC.2017.7858411 (cited on pages 168, 169, 172, 205).

[122]   A. Kostrzewa. "Achieving Performance and Safety in Networks-On-Chip for Real-Time Systems". PhD thesis. TU Braunschweig, 2018 (cited on pages 16, 118, 119, 121–124, 130, 132, 136–139, 142, 151, 152, 154, 155, 160, 161, 163, 164, 167, 168, 187, 189, 197, 238).

[123]   A. Kostrzewa, T. Kadeed, B. Nikolic, and R. Ernst. "Supporting Dynamic Voltage and Frequency Scaling in Networks-On-Chip for Hard Real-Time Systems". In: *24th IEEE International Conference on Embedded and Real-Time Computing Systems and Applications (RTCSA)*. OUTSTANDING PAPER AWARD, BEST PRESENTATION AWARD. Hakodate, Japan, Aug. 2018 (cited on page 136).

[124]   A. Kostrzewa, S. Tobuschat, and R. Ernst. "Self-Aware Network-On-Chip Control in Real-Time Systems". In: *IEEE Design & Test* 35.5 (Oct. 2018), pages 19–27. ISSN: 2168-2356. DOI: 10.1109/MDAT.2017.2763598 (cited on page 205).

[125]   A. Kostrzewa, S. Tobuschat, and R. Ernst. *Technical Report: Resource Manager and Control Layer, Basic Concepts*. Technical report. Institut für Datentechnik und Kommunikationsnetze, Feb. 2018 (cited on pages 16, 79, 105, 118, 119, 121–125, 130, 132, 136–138, 151, 152, 154, 155, 160, 161, 163, 164, 189, 197, 206, 238).

[126]   O. Kotaba, J. Nowotsch, M. Paulitsch, S. M. Petters, and H. Theiling. "Multicore in real-time systems–temporal isolation challenges due to shared resources". In: *Workshop on Industry-Driven Approaches for Cost-effective Certification of Safety-Critical, Mixed-Criticality Systems*. 2014 (cited on page 5).

[127]   R. Kourdy, S. Yazdanpanah, and M. R. N. Rad. "Using the NS-2 Network Simulator for Evaluating Multi Protocol Label Switching in Network-on-Chip". In: *2010 Second International Conference on Computer Research and Development*. May 2010, pages 795–799. DOI: 10.1109/ICCRD.2010.145 (cited on page 42).

[128]   T. Kranich and M. Berekovic. "NoC Switch with Credit Based Guaranteed Service Support Qualified for GALS Systems". In: *Digital System Design: Architectures, Methods and Tools (DSD), 2010 13th Euromicro Conference on*. 2010, pages 53–59. DOI: 10.1109/DSD.2010.30 (cited on pages 32, 143).

[129]   D. Kreutz, F. M. V. Ramos, P. E. Veríssimo, C. E. Rothenberg, S. Azodolmolky, and S. Uhlig. "Software-Defined Networking: A Comprehensive Survey". In: *Proceedings of the IEEE* 103.1 (Jan. 2015), pages 14–76. ISSN: 0018-9219. DOI: 10.1109/JPROC.2014.2371999 (cited on page 121).

[130] A. Kumar, L.-S. Peh, and N. K. Jha. "Token Flow Control". In: *Proceedings of the 41ˢᵗ Annual IEEE/ACM International Symposium on Microarchitecture*. MICRO 41. Washington, DC, USA: IEEE Computer Society, 2008, pages 342–353. ISBN: 978-1-4244-2836-6. DOI: 10.1109/MICRO.2008. 4771803 (cited on page 85).

[131] S. Kunzli, F. Poletti, L. Benini, and L. Thiele. "Combining Simulation and Formal Methods for System-Level Performance Analysis". In: *Proceedings of the Design Automation Test in Europe Conference*. Volume 1. Mar. 2006, pages 1–6. DOI: 10.1109/DATE.2006.244109 (cited on page 42).

[132] S. Kundu and S. Chattopadhyay. *Network-on-chip: the next generation of system-on-chip integration*. Hoboken, NJ: CRC Press, 2014. URL: https: //cds.cern.ch/record/2044063 (cited on page 137).

[133] K. Lampka, K. Huang, and J. Chen. "Dynamic counters and the efficient and effective online power management of embedded real-time systems". In: *2011 Proceedings of the Ninth IEEE/ACM/IFIP International Conference on Hardware/Software Codesign and System Synthesis (CODES+ISSS)*. Oct. 2011, pages 267–276. DOI: 10.1145/2039370.2039412 (cited on page 159).

[134] J.-Y. Le Boudec and P. Thiran. *Network Calculus: A Theory of Deterministic Queuing Systems for the Internet*. Berlin, Heidelberg: Springer-Verlag, 2001. ISBN: 3-540-42184-X (cited on pages 7, 33, 49, 76, 105, 122).

[135] J.-J. Lecler and G. Baillieu. "Application driven network-on-chip architecture exploration & refinement for a complex SoC". In: *Design Automation for Embedded Systems* 15.2 (June 2011), pages 133–158. ISSN: 1572-8080. DOI: 10.1007/s10617-011-9075-5. URL: https://doi.org/10. 1007/s10617-011-9075-5 (cited on page 31).

[136] H. G. Lee, N. Chang, U. Y. Ogras, and R. Marculescu. "On-chip Communication Architecture Exploration: A Quantitative Evaluation of Point-to-point, Bus, and Network-on-chip Approaches". In: *ACM Trans. Des. Autom. Electron. Syst.* 12.3 (May 2008), 23:1–23:20. ISSN: 1084-4309. DOI: 10.1145/1255456.1255460. URL: http://doi.acm.org/10.1145/ 1255456.1255460 (cited on page 42).

[137] J. W. Lee, M. C. Ng, and K. Asanovic. "Globally-Synchronized Frames for Guaranteed Quality-of-Service in On-Chip Networks". In: *Proceedings of the 35ᵗʰ Annual International Symposium on Computer Architecture*. ISCA '08. Washington, DC, USA: IEEE Computer Society, 2008, pages 89–100. ISBN: 978-0-7695-3174-8. DOI: 10.1109/ISCA.2008.31 (cited on page 82).

[138]   P. Lekkas. *Network Processors: Architectures, Protocols and Platforms.*
        Telecom Engineering. McGraw-Hill Education, 2003. ISBN: 9780071429122.
        URL: https://books.google.de/books?id=uJ49S774t1IC (cited on
        page 105).

[139]   P. Lieverse, P. Van Der Wolf, K. Vissers, and E. Deprettere. "A Methodol-
        ogy for Architecture Exploration of Heterogeneous Signal Processing Sys-
        tems". In: *Journal of VLSI signal processing systems for signal, image and
        video technology* 29.3 (Nov. 2001), pages 197–207. ISSN: 0922-5773. DOI:
        10.1023/A:1012231429554. URL: https://doi.org/10.1023/A:
        1012231429554 (cited on pages 46, 47).

[140]   P. Lieverse, P. van der Wolf, E. Deprettere, and K. Vissers. "A methodology
        for architecture exploration of heterogeneous signal processing systems". In:
        *1999 IEEE Workshop on Signal Processing Systems. SiPS 99. Design and
        Implementation (Cat. No.99TH8461).* 1999, pages 181–190. DOI: 10.1109/
        SIPS.1999.822323 (cited on page 46).

[141]   M. Lis, K. Sup, S. Myong, H. Cho, P. Ren, O. Khan, and S. Devadas.
        "DARSIM: a parallel cycle-level NoC simulator". In: *in Sixth Workshop on
        Modeling, Benchmarking, and Simulation (MoBS.* 2010 (cited on page 42).

[142]   J. W. Liu. *Real-time systems.* Prentice Hall, 2000. ISBN: 978-0-13-099651-0
        (cited on pages 9, 10).

[143]   Z. Lu, A. Jantsch, and I. Sander. "Feasibility analysis of messages for on-chip
        networks using wormhole routing". In: *Proceedings of the ASP-DAC 2005.
        Asia and South Pacific Design Automation Conference, 2005.* Volume 2. Jan.
        2005, 960–964 Vol. 2. DOI: 10.1109/ASPDAC.2005.1466499 (cited on
        page 48).

[144]   Z. Lu, R. Thid, M. Millberg, E. Nilsson, and A. Jantsch. "NNSE: Nostrum
        Network-on-Chip Simulation Environment". In: *Proceedings of Swedish
        System-on-Chip Conference, Stockholm, Sweden, April 2005.* QC 20100524.
        2005. URL: http://www.edacentrum.de/veranstaltungen/2005/
        date05/ubooth/descriptions/Description_sw_nnse.pdf (cited on
        page 42).

[145]   S. Mahadevan, F. Angiolini, M. Storoaard, R. G. Olsen, J. Sparso, and J.
        Madsen. "Network traffic generator model for fast network-on-chip simula-
        tion". In: *Design, Automation and Test in Europe.* Mar. 2005, 780–785 Vol.
        2. DOI: 10.1109/DATE.2005.22 (cited on page 41).

[146]   T. Marescaux and H. Corporaal. "Introducing the SuperGT Network-on-
        Chip; SuperGT QoS: more than just GT". In: *2007 44th ACM/IEEE Design
        Automation Conference.* June 2007, pages 116–121 (cited on page 105).

[147]   C. Meinel and H. Sack. *Internetworking - Technische Grundlagen und An-
        wendungen.* Jan. 2011. ISBN: 978-3-540-92939-0 (cited on page 105).

[148] A. V. de Mello, L. C. Ost, F. G. Moraes, and N. L. V. Calazans. "Evaluation of routing algorithms on mesh based nocs". In: *PUCRS, Av. Ipiranga* (2004) (cited on page 22).

[149] M. Millberg, E. Nilsson, R. Thid, and A. Jantsch. "Guaranteed bandwidth using looped containers in temporally disjoint networks within the nostrum network on chip". In: *Design, Automation and Test in Europe Conference and Exhibition, 2004. Proceedings*. Volume 2. Feb. 2004, pages 890–8952. DOI: 10.1109/DATE.2004.1269001 (cited on pages 34, 81).

[150] M. Millberg, E. Nilsson, R. Thid, S. Kumar, and A. Jantsch. "The Nostrum backbone-a communication protocol stack for Networks on Chip". In: *17th International Conference on VLSI Design. Proceedings*. 2004, pages 693–696. DOI: 10.1109/ICVD.2004.1261005 (cited on page 34).

[151] M. Millberg, E. Nilsson, R. Thid, S. Kumar, and A. Jantsch. "The Nostrum backbone-a communication protocol stack for networks on chip". In: *VLSI Design, 2004. Proceedings. 17th International Conference on*. IEEE. 2004, pages 693–696 (cited on page 25).

[152] I. Miro Panades, A. Greiner, and A. Sheibanyrad. "A Low Cost Network-on-Chip with Guaranteed Service Well Suited to the GALS Approach". In: *Nano-Networks and Workshops, 2006. NanoNet '06. 1st International Conference on*. Sept. 2006, pages 1–5. DOI: 10.1109/NANONET.2006.346219 (cited on page 81).

[153] B. Motruk, J. Diemer, R. Buchty, R. Ernst, and M. Berekovic. "IDAMC: A Many-Core Platform with Run-Time Monitoring for Mixed-Criticality". In: *High-Assurance Systems Engineering (HASE), 2012 IEEE 14th International Symposium on*. 2012, pages 24–31. DOI: 10.1109/HASE.2012.19 (cited on page 158).

[154] S. Mubeen. *Evaluation of Source Routing for Mesh Topology Network on Chip Platforms*. 2009 (cited on page 147).

[155] S. Mubeen and S. Kumar. "Designing Efficient Source Routing for Mesh Topology Network on Chip Platforms". In: *2010 13th Euromicro Conference on Digital System Design: Architectures, Methods and Tools*. Sept. 2010, pages 181–188. DOI: 10.1109/DSD.2010.57 (cited on pages 22, 147).

[156] N. Muralimanohar and R. Balasubramonian. "Interconnect Design Considerations for Large NUCA Caches". In: *Proceedings of the 34th Annual International Symposium on Computer Architecture*. ISCA '07. San Diego, California, USA: ACM, 2007, pages 369–380. ISBN: 978-1-59593-706-3. DOI: 10.1145/1250662.1250708 (cited on page 81).

[157]   M. Negrean, M. Neukirchner, S. Stein, S. Schliecker, and R. Ernst. "Bound-
        ing Mode Change Transition Latencies for Multi-Mode Real-Time Dis-
        tributed Applications". In: *16th IEEE International Conference on Emerging
        Technologies and Factory Automation (ETFAÍ1)*. Toulouse, France, Sept.
        2011. URL: http://dx.doi.org/10.1109/ETFA.2011.6059009 (cited
        on page 172).

[158]   M. Negrean, R. Ernst, and S. Schliecker. "Mastering Timing Challenges for
        the Design of Multi-Mode Applications on Multi-Core Real-Time Embedded
        Systems". In: *6th International Congress on Embedded Real-Time Software
        and Systems (ERTS)*. Toulouse, France, Feb. 2012 (cited on page 172).

[159]   M. F. Negrean. "Performance Analysis of Multi-Core Multi-Mode Systems
        with Shared Resources - Principles and Application to AUTOSAR -". PhD
        thesis. TUBS, 2015 (cited on page 172).

[160]   M. Neukirchner, T. Michaels, P. Axer, S. Quinton, and R. Ernst. "Monitoring
        Arbitrary Activation Patterns in Real-Time Systems". In: *Real-Time Systems
        Symposium (RTSS), 2012 IEEE 33rd*. Dec. 2012, pages 293–302. DOI: 10.
        1109/RTSS.2012.80 (cited on page 159).

[161]   M. Neukirchner, K. Lampka, S. Quinton, and R. Ernst. "Multi-mode monitor-
        ing for mixed-criticality real-time systems". In: *2013 International Confer-
        ence on Hardware/Software Codesign and System Synthesis (CODES+ISSS)*.
        Sept. 2013, pages 1–10. DOI: 10.1109/CODES-ISSS.2013.6659021
        (cited on page 159).

[162]   B. Nikolić, S. Tobuschat, L. Soares Indrusiak, R. Ernst, and A. Burns. "Real-
        time analysis of priority-preemptive NoCs with arbitrary buffer sizes and
        router delays". In: *Real-Time Systems* (June 2018). ISSN: 1573-1383. DOI:
        10.1007/s11241-018-9312-0. URL: https://doi.org/10.1007/
        s11241-018-9312-0 (cited on pages 49, 206).

[163]   U. Y. Ogras and R. Marculescu. ""It's a small world after all": NoC perfor-
        mance optimization via long-range link insertion". In: *IEEE Transactions on
        Very Large Scale Integration (VLSI) Systems* 14.7 (July 2006), pages 693–
        706. ISSN: 1063-8210. DOI: 10.1109/TVLSI.2006.878263 (cited on
        page 42).

[164]   U. Y. Ogras and R. Marculescu. "Modeling, Analysis and Optimization
        of Network-on-Chip Communication Architectures". In: Lecture Notes In
        Electrical Engineering, 2013 (cited on pages 18, 47).

[165]   A. Padovitz, S. W. Loke, and A. Zaslavsky. "Awareness and Agility for
        Autonomic Distributed Systems: Platform-Independent Publish-Subscribe
        Event-Based Communication for Mobile Agents". In: *14th International
        Workshop on Database and Expert Systems Applications, 2003. Proceed-
        ings.(DEXA)*. Volume 00. Sept. 2003, page 669. DOI: 10.1109/DEXA.

2003. 1232098. URL: doi.ieeecomputersociety.org/10.1109/
DEXA.2003.1232098 (cited on page 199).

[166] P. P. Pande, C. Grecu, M. Jones, A. Ivanov, and R. Saleh. "Performance eval-
uation and design trade-offs for network-on-chip interconnect architectures".
In: *IEEE Transactions on Computers* 54.8 (Aug. 2005), pages 1025–1040.
ISSN: 0018-9340. DOI: 10.1109/TC.2005.134 (cited on page 145).

[167] R. Parikh and V. Bertacco. "Formally Enhanced Runtime Verification to Ensure NoC Functional Correctness". In: *Proceedings of the 44th Annual
IEEE/ACM International Symposium on Microarchitecture*. MICRO-44.
Porto Alegre, Brazil: ACM, 2011, pages 410–419. ISBN: 978-1-4503-1053-
6. DOI: 10.1145/2155620.2155668. URL: http://doi.acm.org/10.
1145/2155620.2155668 (cited on page 201).

[168] R. Parikh and V. Bertacco. "ForEVeR: A Complementary Formal and Run-
time Verification Approach to Correct NoC Functionality". In: *ACM Trans.
Embed. Comput. Syst.* 13.3s (Mar. 2014), 104:1–104:30. ISSN: 1539-9087.
DOI: 10.1145/2514871. URL: http://doi.acm.org/10.1145/
2514871 (cited on page 201).

[169] G. Passas, M. Katevenis, and D. Pnevmatikatos. "A 128 x 128 x 24Gb/s
Crossbar Interconnecting 128 Tiles in a Single Hop and Occupying 6% of
Their Area". In: *Proceedings of the 2010 Fourth ACM/IEEE International
Symposium on Networks-on-Chip*. NOCS '10. Washington, DC, USA: IEEE
Computer Society, 2010, pages 87–95. ISBN: 978-0-7695-4053-5. DOI: 10.
1109/NOCS.2010.37. URL: http://dx.doi.org/10.1109/NOCS.
2010.37 (cited on page 18).

[170] G. Passas, M. Katevenis, and D. N. Pnevmatikatos. "Crossbar NoCs Are
Scalable Beyond 100 Nodes". In: *IEEE Transactions on Computer-Aided
Design of Integrated Circuits and Systems* 31 (2012), pages 573–585 (cited
on page 18).

[171] R. Pellizzoni, P. Meredith, M.-Y. Nam, M. Sun, M. Caccamo, and L. Sha.
"Handling Mixed-criticality in SoC-based Real-time Embedded Systems".
In: *EMSOFT*. ACM, 2009 (cited on page 167).

[172] R. Pellizzoni and et al. "A predictable execution model for cots-based embedded systems". In: *RTAS*. 2011 (cited on page 167).

[173] S. G. Pestana, E. Rijpkema, A. Radulescu, K. Goossens, and O. P. Gangwal.
"Cost-performance trade-offs in networks on chip: a simulation-based approach". In: *Proceedings Design, Automation and Test in Europe Conference
and Exhibition*. Volume 2. Feb. 2004, 764–769 Vol.2. DOI: 10.1109/DATE.
2004.1268972 (cited on page 41).

[174]   A. Psarras, I. Seitanidis, C. Nicopoulos, and G. Dimitrakopoulos. "PhaseNoC: TDM Scheduling at the Virtual-channel Level for Efficient Network Traffic Isolation". In: *Proceedings of the 2015 Design, Automation & Test in Europe Conference & Exhibition.* DATE '15. Grenoble, France: EDA Consortium, 2015, pages 1090–1095. ISBN: 978-3-9815370-4-8. URL: http://dl.acm.org/citation.cfm?id=2755753.2757066 (cited on pages 81, 122).

[175]   V. Puente, J. A. Gregorio, and R. Beivide. "SICOSYS: an integrated framework for studying interconnection network performance in multiprocessor systems". In: *Proceedings 10th Euromicro Workshop on Parallel, Distributed and Network-based Processing.* 2002, pages 15–22. DOI: 10.1109/EMPDP.2002.994207 (cited on page 42).

[176]   V. Puente, J. Gregorio, C. Izu, and R. Beivide. "Impact of the Head-of-Line Blocking on Parallel Computer Networks: Hardware to Applications". In: *In European Conference on Parallel Processing.* 1999, pages 1222–1230 (cited on page 119).

[177]   Y. Qian, Z. Lu, and W. Dou. "Analysis of worst-case delay bounds for best-effort communication in wormhole networks on chip". In: *Networks-on-Chip, 2009. NoCS 2009. 3rd ACM/IEEE International Symposium on.* May 2009, pages 44–53. DOI: 10.1109/NOCS.2009.5071444 (cited on pages 49, 105).

[178]   Y. Qian, Z. Lu, and W. Dou. "Analysis of communication delay bounds for network on chips". In: *2009 Asia and South Pacific Design Automation Conference.* Jan. 2009, pages 7–12. DOI: 10.1109/ASPDAC.2009.4796433 (cited on page 48).

[179]   Z. Qian, D. C. Juan, P. Bogdan, C. Y. Tsui, D. Marculescu, and R. Marculescu. "A comprehensive and accurate latency model for Network-on-Chip performance analysis". In: *2014 19th Asia and South Pacific Design Automation Conference (ASP-DAC).* Jan. 2014, pages 323–328. DOI: 10.1109/ASPDAC.2014.6742910 (cited on page 49).

[180]   D. Rahmati, S. Murali, L. Benini, F. Angiolini, G. D. Micheli, and H. Sarbazi-Azad. "A method for calculating hard QoS guarantees for Networks-on-Chip". In: *2009 IEEE/ACM International Conference on Computer-Aided Design - Digest of Technical Papers.* Nov. 2009, pages 579–586. DOI: 10.1145/1687399.1687507 (cited on page 48).

[181]   E. A. Rambo, A. Tschiene, J. Diemer, L. Ahrendts, and R. Ernst. "Failure Analysis of a Network-on-Chip for Real-Time Mixed-Critical Systems". In: *Design, Automation & Test in Europe Conference & Exhibition (DATE), 2014.* Mar. 2014, pages 1–4. URL: http://dx.doi.org/10.7873/DATE.2014.288 (cited on pages 148, 201).

[182] E. A. Rambo, A. Tschiene, J. Diemer, L. Ahrendts, and R. Ernst. "FMEA-Based Analysis of a Network-on-Chip for Mixed-Critical Systems". In: *Networks on Chip (NoCS), 2014 Eighth IEEE/ACM International Symposium on*. 2014. URL: http://dx.doi.org/10.1109/NOCS.2014.7008759 (cited on pages 148, 201).

[183] E. A. Rambo and R. Ernst. "Worst-case Communication Time Analysis of Networks-on-chip with Shared Virtual Channels". In: *Proceedings of the 2015 Design, Automation & Test in Europe Conference & Exhibition*. DATE '15. Grenoble, France: EDA Consortium, 2015, pages 537–542. ISBN: 978-3-9815370-4-8. URL: http://dl.acm.org/citation.cfm?id=2755753.2755874 (cited on pages 48, 52, 68, 70, 74, 75, 85).

[184] E. A. Rambo and R. Ernst. "Providing Flexible and Reliable on-Chip Network Communication with Real-Time Constraints". In: *1st International Workshop on Resiliency in Embedded Electronic Systems (REES)*. Amsterdam, Netherlands, Oct. 2015 (cited on pages 32, 148).

[185] E. A. Rambo, C. Seitz, S. Saidi, and R. Ernst. "Bridging the Gap between Resilient Networks-on-Chip and Real-Time Systems". In: *IEEE Transactions on Emerging Topics in Computing* PP.99 (Aug. 2017). URL: https://doi.org/10.1109/TETC.2017.2736783 (cited on page 148).

[186] E. A. Rambo, C. Seitz, S. Saidi, and R. Ernst. "Designing Networks-on-Chip for High Assurance Real-Time Systems". In: *Pacific Rim International Symposium on Dependable Computing (PRDC), 2017*. Christchurch, New Zealand, Jan. 2017. URL: https://doi.org/10.1109/PRDC.2017.32 (cited on pages 143, 148).

[187] V. Rantala, T. Lehtonen, and J. Plosila. *Network on Chip Routing Algorithms*. TUCS technical report. Turku Centre for Computer Science, 2006. ISBN: 9789521217647. URL: http://books.google.de/books?id=0irQMgAACAAJ (cited on page 22).

[188] K. Richter. "Compositional scheduling analysis using standard event models: the SymTA/S approach". PhD thesis. 2005, pages 1–204 (cited on page 50).

[189] I. Saastamoinen, M. Alho, and J. Nurmi. "Buffer implementation for Proteo network-on-chip". In: *Circuits and Systems, 2003. ISCAS '03. Proceedings of the 2003 International Symposium on*. Volume 2. May 2003, II-113-II-116 vol.2. DOI: 10.1109/ISCAS.2003.1205906 (cited on page 34).

[190] S. Saidi, R. Ernst, S. Uhrig, H. Theiling, and B. D. de Dinechin. "The shift to multicores in real-time and safety-critical systems". In: *2015 International Conference on Hardware/Software Codesign and System Synthesis (CODES+ISSS)*. Oct. 2015, pages 220–229. DOI: 10.1109/CODESISSS.2015.7331385 (cited on page 33).

[191]   E. Salminen, T. Kangas, J. Riihimäki, V. Lahtinen, K. Kuusilinna, and
        T. D. Hämäläinen. "Benchmarking Mesh and Hierarchical Bus Networks in
        System-on-Chip Context". In: *Embedded Computer Systems: Architectures,
        Modeling, and Simulation*. Edited by T. D. Hämäläinen, A. D. Pimentel,
        J. Takala, and S. Vassiliadis. Berlin, Heidelberg: Springer Berlin Heidelberg,
        2005, pages 354–363. ISBN: 978-3-540-31664-0 (cited on page 41).

[192]   E. Salminen, T. Kangas, V. Lahtinen, J. Riihimäki, K. Kuusilinna, and
        T. D. Hämäläinen. "Benchmarking Mesh and Hierarchical Bus Networks in
        System-on-chip Context". In: *J. Syst. Archit.* 53.8 (Aug. 2007), pages 477–
        488. ISSN: 1383-7621. DOI: 10.1016/j.sysarc.2006.11.006. URL:
        http://dx.doi.org/10.1016/j.sysarc.2006.11.006 (cited on
        page 41).

[193]   L. Schenato, B. Sinopoli, M. Franceschetti, K. Poolla, and S. Sastry. "Foun-
        dations of Control and Estimation Over Lossy Networks". In: *Proceed-
        ings of the IEEE* 95.1 (Jan. 2007), pages 163–187. ISSN: 0018-9219. DOI:
        10.1109/JPROC.2006.887306 (cited on page 10).

[194]   S. Schliecker, J. Rox, M. Ivers, and R. Ernst. "Providing Accurate Event
        Models for the Analysis of Heterogeneous Multiprocessor Systems". In:
        *Proceedings of the 6th IEEE/ACM/IFIP International Conference on Hard-
        ware/Software Codesign and System Synthesis*. CODES+ISSS '08. Atlanta,
        GA, USA: ACM, 2008, pages 185–190. ISBN: 978-1-60558-470-6. DOI:
        10.1145/1450135.1450177 (cited on page 50).

[195]   J. Schlatow, M. Mostl, S. Tobuschat, T. Ishigooka, and R. Ernst. "Data-Age
        Analysis and Optimisation for Cause-Effect Chains in Automotive Control
        Systems". In: *2018 IEEE 13th International Symposium on Industrial Em-
        bedded Systems (SIES)*. June 2018, pages 1–9. DOI: 10.1109/SIES.2018.
        8442077 (cited on page 207).

[196]   C. Seiculescu, S. Murali, L. Benini, and G. D. Micheli. "SunFloor 3D: A
        tool for Networks On Chip topology synthesis for 3D systems on chips". In:
        2009 Design, Automation Test in Europe Conference Exhibition. Apr. 2009,
        pages 9–14. DOI: 10.1109/DATE.2009.5090625 (cited on page 42).

[197]   C. Seitz. "Evaluation, Implementation and Integration of a Resilient Network-
        on-Chip Architecture for the IDAMC platform". Master's thesis. TU Braun-
        schweig, Jan. 2016 (cited on pages 148, 152).

[198]   Z. Shi and A. Burns. "Real-Time Communication Analysis for On-Chip
        Networks with Wormhole Switching". In: *Networks-on-Chip, 2008. NoCS
        2008. Second ACM/IEEE International Symposium on*. Apr. 2008, pages 161–
        170. DOI: 10.1109/NOCS.2008.4492735 (cited on pages 41, 48, 52).

[199] Z. Shi and A. Burns. "Real-Time Communication Analysis with a Priority Share Policy in On-Chip Networks". In: *2009 21st Euromicro Conference on Real-Time Systems*. July 2009, pages 3–12. DOI: 10.1109/ECRTS.2009.17 (cited on page 48).

[200] Z. Shi and A. Burns. "Improvement of schedulability analysis with a priority share policy in on-chip networks". In: *17th International Conference on Real-Time and Network Systems*. 2009, pages 75–84 (cited on page 48).

[201] J. A. Stankovic, K. Ramamritham, and M. Spuri. *Deadline Scheduling for Real-Time Systems: Edf and Related Algorithms*. Norwell, MA, USA: Kluwer Academic Publishers, 1998. ISBN: 0792382692 (cited on pages 10, 15, 80, 84, 141, 172).

[202] S. Stein, J. Diemer, M. Ivers, S. Schliecker, and R. Ernst. *On the Convergence of the SymTA/S analysis*. Technical report. Braunschweig, Germany: TU Braunschweig, Nov. 2008 (cited on page 70).

[203] M. Stępniewska, O. Stankiewicz, A. Łuczak, and J. Siast. "Embedded debugging for NoCs". In: *Proceedings of the 17th International Conference Mixed Design of Integrated Circuits and Systems - MIXDES 2010*. June 2010, pages 601–606 (cited on page 202).

[204] S. Stein. "Allowing Flexibility in Critical Systems: The EPOC Framework". PhD thesis. TU BS, 2012 (cited on pages 70, 71).

[205] STMicroelectronics. *STNoC: Building a New System-on-Chip Paradigm*. *Whitepaper.* 2005. URL: www.st.com (cited on page 36).

[206] D. Stöhrmann. "Implementierung, Integration und Test von prioritätsbasierten Routern im IDA NoC". Master's thesis. TU Braunschweig, März 2017 (cited on pages 143, 145).

[207] A. Tanenbaum. *Computer networks*. Upper Saddle River, NJ: Prentice Hall PTR, 2003. ISBN: 0130661023 (cited on page 152).

[208] S. Tang and Q. Xu. "A Multi-Core Debug Platform for NoC-Based Systems". In: *2007 Design, Automation Test in Europe Conference Exhibition*. Apr. 2007, pages 1–6. DOI: 10.1109/DATE.2007.364402 (cited on page 202).

[209] P. P. Tang and T. .-.-. C. Tai. "Network traffic characterization using token bucket model". In: *IEEE INFOCOM '99. Conference on Computer Communications. Proceedings. Eighteenth Annual Joint Conference of the IEEE Computer and Communications Societies. The Future is Now (Cat. No.99CH36320)*. Volume 1. Mar. 1999, 51–62 vol.1. DOI: 10.1109/INFCOM.1999.749252 (cited on page 105).

[210]   M. B. Taylor, J. Kim, J. Miller, D. Wentzlaff, F. Ghodrat, B. Greenwald,
        H. Hoffman, P. Johnson, J.-W. Lee, W. Lee, A. Ma, A. Saraf, M. Seneski,
        N. Shnidman, V. Strumpen, M. Frank, S. Amarasinghe, and A. Agarwal.
        "The Raw microprocessor: a computational fabric for software circuits and
        general-purpose programs". In: *IEEE Micro* 22.2 (Mar. 2002), pages 25–35.
        ISSN: 0272-1732. DOI: 10.1109/MM.2002.997877 (cited on page 35).

[211]   M. B. Taylor, J. Psota, A. Saraf, N. Shnidman, V. Strumpen, M. Frank,
        S. Amarasinghe, A. Agarwal, W. Lee, J. Miller, D. Wentzlaff, I. Bratt, B.
        Greenwald, H. Hoffmann, P. Johnson, and J. Kim. "Evaluation of the Raw
        microprocessor: an exposed-wire-delay architecture for ILP and streams". In:
        *Proceedings. 31st Annual International Symposium on Computer Architec-
        ture, 2004*. June 2004, pages 2–13. DOI: 10.1109/ISCA.2004.1310759
        (cited on page 35).

[212]   C. Temple. "Avoiding the babbling-idiot failure in a time-triggered commu-
        nication system". In: *Digest of Papers. Twenty-Eighth Annual International
        Symposium on Fault-Tolerant Computing (Cat. No.98CB36224)*. June 1998,
        pages 218–227. DOI: 10.1109/FTCS.1998.689473 (cited on page 237).

[213]   L. Thiele, S. Chakraborty, and M. Naedele. "Real-time calculus for schedul-
        ing hard real-time systems". In: *Circuits and Systems, 2000. Proceedings.
        ISCAS 2000 Geneva. The 2000 IEEE International Symposium on*. Volume 4.
        2000, pages 101–1044. DOI: 10.1109/ISCAS.2000.858698 (cited on
        pages 7, 52).

[214]   Tilera. *Tile Processor Architecture Overview for the TILEPro Series, UG120*.
        1.2. Tilera Corporation. Feb. 2013. URL: https://www.mellanox.com/
        repository/solutions/tile-scm/docs/UG120-Architecture-
        Overview-TILEPro.pdf (cited on pages 36, 78, 121, 122, 142).

[215]   K. W. Tindell, A. Burns, and A. J. Wellings. "An Extendible Approach for
        Analyzing Fixed Priority Hard Real-time Tasks". In: *Real-Time Syst.* 6.2
        (Mar. 1994), pages 133–151. ISSN: 0922-6443. DOI: 10.1007/BF01088593
        (cited on pages 51, 52).

[216]   S. Tobuschat, P. Axer, R. Ernst, and J. Diemer. "IDAMC: A NoC for mixed
        criticality systems". In: *2013 IEEE 19th International Conference on Em-
        bedded and Real-Time Computing Systems and Applications*. Aug. 2013,
        pages 149–156. DOI: 10.1109/RTCSA.2013.6732214 (cited on pages 5,
        32, 85, 135, 143, 158, 186, 203).

[217]   S. Tobuschat, M. Neukirchner, L. Ecco, and R. Ernst. "Workload-aware shap-
        ing of shared resource accesses in mixed-criticality systems". In: *Hardware/-
        Software Codesign and System Synthesis (CODES+ISSS), 2014 International
        Conference on*. Oct. 2014, pages 1–10. DOI: 10.1145/2656075.2656105
        (cited on pages 45, 81, 84, 86, 137, 141, 159, 172, 203).

[218] S. Tobuschat, R. Ernst, A. Hamann, and D. Ziegenbein. "System-level timing feasibility test for cyber-physical automotive systems". In: *2016 11th IEEE Symposium on Industrial Embedded Systems (SIES)*. May 2016, pages 1–10. DOI: 10.1109/SIES.2016.7509419 (cited on pages 10, 206).

[219] S. Tobuschat and R. Ernst. "Efficient Latency Guarantees for Mixed-Criticality Networks-on-Chip". In: *2017 IEEE Real-Time and Embedded Technology and Applications Symposium (RTAS)*. Apr. 2017, pages 113–122. DOI: 10.1109/RTAS.2017.31 (cited on pages 79, 84, 204).

[220] S. Tobuschat and R. Ernst. "Real-time communication analysis for Networks-on-Chip with backpressure". In: *Design, Automation Test in Europe Conference Exhibition (DATE), 2017*. Mar. 2017, pages 590–595. DOI: 10.23919/DATE.2017.7927055 (cited on pages 45, 52, 54, 68, 105, 204).

[221] S. Tobuschat and R. Ernst. "Providing Throughput Guarantees in Mixed-criticality Networks-on-Chip". In: *2017 30th IEEE International System-on-Chip Conference (SOCC) (SOCC 2017)*. Munich, Germany, Sept. 2017, pages 207–212 (cited on pages 79, 204).

[222] S. Tobuschat, A. Kostrzewa, F. K. Bapp, and C. Dropmann. "Online monitoring for safety-critical multicore systems". In: *it - Information Technology* (2017). URL: https://doi.org/10.1515/itit-2017-0028 (cited on page 207).

[223] S. Tobuschat, A. Kostrzewa, and R. Ernst. "Selective congestion control for mixed-critical networks-on-chip". In: *Integration, the {VLSI} Journal* (2017), pages -. ISSN: 0167-9260. DOI: https://doi.org/10.1016/j.vlsi.2017.12.003. URL: https://www.sciencedirect.com/science/article/pii/S0167926017302341 (cited on pages 16, 79, 105, 118, 119, 121, 123–125, 132, 135–138, 151, 152, 154, 155, 160, 163, 164, 167, 168, 172, 197, 205).

[224] W. C. Tsai, H. E. Lin, Y. C. Lan, S. J. Chen, and Y. H. Hu. "A novel flow fluidity meter for BiNoC bandwidth resource allocation". In: *2015 28th IEEE International System-on-Chip Conference (SOCC)*. Sept. 2015, pages 281–286. DOI: 10.1109/SOCC.2015.7406964 (cited on page 84).

[225] S. R. Vangal, J. Howard, G. Ruhl, S. Dighe, H. Wilson, J. Tschanz, D. Finan, A. Singh, T. Jacob, S. Jain, V. Erraguntla, C. Roberts, Y. Hoskote, N. Borkar, and S. Borkar. "An 80-Tile Sub-100-W TeraFLOPS Processor in 65-nm CMOS". In: *IEEE Journal of Solid-State Circuits* 43.1 (Jan. 2008), pages 29–41. ISSN: 0018-9200. DOI: 10.1109/JSSC.2007.910957 (cited on page 140).

[226]   A. Varga and R. Hornig. "An Overview of the OmNet++ Simulation En-
         vironment". In: *Proceedings of the 1st International Conference on Simu-
         lation Tools and Techniques for Communications, Networks and Systems
         & Workshops*. Simutools '08. Marseille, France: ICST (Institute for Com-
         puter Sciences, Social-Informatics and Telecommunications Engineering),
         2008, 60:1–60:10. ISBN: 978-963-9799-20-2. URL: http://dl.acm.org/
         citation.cfm?id=1416222.1416290 (cited on pages 42, 72, 167).

[227]   A. Varga. "OmNet++". In: *Modeling and Tools for Network Simulation*.
         Edited by K. Wehrle, M. Güneş, and J. Gross. Berlin, Heidelberg: Springer
         Berlin Heidelberg, 2010, pages 35–59. ISBN: 978-3-642-12331-3. DOI: 10.
         1007/978-3-642-12331-3_3. URL: https://doi.org/10.1007/978-
         3-642-12331-3_3 (cited on pages 42, 72, 167).

[228]   B. Vermeulen and K. Goossens. "A Network-on-Chip monitoring infrastruc-
         ture for communication-centric debug of embedded multi-processor SoCs".
         In: *2009 International Symposium on VLSI Design, Automation and Test*.
         Apr. 2009, pages 183–186. DOI: 10.1109/VDAT.2009.5158125 (cited on
         page 202).

[229]   P. W. Viglucci and A. Carpenter. "ENoCS: An Interactive Educational
         Network-on-Chip Simulator". In: *ASEE Annual Conference & Exposition*.
         2016 (cited on page 42).

[230]   J.-D. Völkel. "Implementierung von Control-Plane im IDA Network-On-
         Chip". Master's thesis. TUBS, July 2018 (cited on pages 161, 167, 187).

[231]   E. Wandeler and L. Thiele. "Real-time Interfaces for Interface-based Design
         of Real-time Systems with Fixed Priority Scheduling". In: *Proceedings of
         the 5th ACM International Conference on Embedded Software*. EMSOFT
         '05. Jersey City, NJ, USA: ACM, 2005, pages 80–89. ISBN: 1-59593-091-4.
         DOI: 10.1145/1086228.1086246 (cited on page 82).

[232]   J. Wang, Y. Li, Q. Peng, and T. Tan. "A dynamic priority arbiter for Network-
         on-Chip". In: *2009 IEEE International Symposium on Industrial Embedded
         Systems*. July 2009, pages 253–256. DOI: 10.1109/SIES.2009.5196222
         (cited on pages 89, 90).

[233]   D. Wentzlaff, P. Griffin, H. Hoffmann, L. Bao, B. Edwards, C. Ramey, M.
         Mattina, C.-C. Miao, J. F. Brown III, and A. Agarwal. "On-Chip Interconnec-
         tion Architecture of the Tile Processor". In: *IEEE Micro* 27.5 (Sept. 2007),
         pages 15–31. ISSN: 0272-1732. DOI: 10.1109/MM.2007.89 (cited on
         pages 36, 49, 54, 78, 121, 122, 140, 142).

[234]   D. Wiklund and D. Liu. "SoCBUS: switched network on chip for hard real
         time embedded systems". In: *IPDPS*. Apr. 2003, 8 pp.-. DOI: 10.1109/
         IPDPS.2003.1213180 (cited on page 35).

[235] D. Wiklund, A. Ehliar, and D. Liu. "Design of an Internet core router using the SoCBUS network on chip". In: *International Symposium on Signals, Circuits and Systems, 2005. ISSCS 2005*. Volume 2. July 2005, 513–516 Vol. 2. DOI: 10.1109/ISSCS.2005.1511290 (cited on page 35).

[236] D. Wiklund. "Development and performance evaluation of networks on chip". PhD thesis. Linköping University, Department of Electrical Engineering, 2005, page 168. ISBN: 91-85297-62-3 (cited on page 35).

[237] Y. J. Yoon, N. Concer, M. Petracca, and L. Carloni. "Virtual Channels vs. Multiple Physical Networks: A Comparative Analysis". In: *Proceedings of the 47th Design Automation Conference*. DAC '10. Anaheim, California: ACM, 2010, pages 162–165. ISBN: 978-1-4503-0002-5. DOI: 10.1145/1837274.1837315. URL: http://doi.acm.org/10.1145/1837274.1837315 (cited on page 142).

[238] A. Zaib, J. Heißwolf, A. Weichslgartner, T. Wild, J. Teich, J. Becker, and A. Herkersdorf. "AUTO-GS: Self-Optimization of NoC Traffic through Hardware Managed Virtual Connections". In: *2013 Euromicro Conference on Digital System Design*. Sept. 2013, pages 761–768. DOI: 10.1109/DSD.2013.87 (cited on pages 123, 172).

[239] D. Ziegenbein and A. Hamann. "Timing-aware Control Software Design for Automotive Systems". In: *DAC*. 2015 (cited on page 10).

[240] H. Zimmer and A. Jantsch. "A fault model notation and error-control scheme for switch-to-switch buses in a network-on-chip". In: *First IEEE/ACM/IFIP International Conference on Hardware/ Software Codesign and Systems Synthesis (IEEE Cat. No.03TH8721)*. Oct. 2003, pages 188–193. DOI: 10.1109/CODESS.2003.1275281 (cited on page 148).

Glossary

adaptability see flexibility.

analysability The capability of a component or system to be diagnosed for deficiencies or causes of failures in the component or system, or for the parts to be modified to be identified. Analysability requires a formal model of the component or system (and, if applicable, of the external stimulus). See also Predictability.

availability The ability to be in a state to perform as required. Availability depends upon the combined characteristics of the reliability, recover-ability, and maintainability of the item, and the maintenance support performance [52].

babbling idiot For interconnects, the babbling-idiot denotes a node that busies the interconnect unduly [43; 212].

best-case responce time (BCRT) Minimum possible time between receiving a stimulus and delivering an appropriate response or reaction. In scope of this thesis, the best-case response time of a flit at a network router describes the minimum possible time between the arrival of the flit at the router and its forwarding including blocking through other flits.

best-effort (BE) Best-effort denotes the property to require no gurantees.

commercial off-the-shelf (COTS) An entitiy (e.g. software or hardware components) defined by market-driven needs, commercially available, and whose fitness for purpose has been demonstrated by a broad spectrum of commercial users [52].

control traffic (CT) Control traffic sent by the resource manager of clients using the resource management (e.g. control layer approacj [122; 125].

deadline The point in time when an execution of an entity must be finished [8].

dependability The ability to perform as and when required. Dependability includes availability, reliability, recoverability, maintainability, and maintenance support performance, and, in some cases, other characteristics such as durability, safety and security. Dependability is used as a collective term for the time-related quality characteristics of an item [52].

dependable See dependability.

distributed traffic shaping (DTS) Distributed Traffic Shaping is a QoS scheme for NoCs [59].

electronic control unit (ECU) Embedded computer system consisting out of at least one CPU and corresponding periphery which is placed in one housing [8].

flexibility Flexibility is related to the effort to adapt the entity/system to system internal and external changing conditions as well to re-use the entity/system for different use cases.

flow control digit (flit) The unit of buffer management and flow control for wormhole flow control [54; 69].

globally asynchronous locally synchronous (GALS) Globally asynchronous locally synchronous is a paradigm for designing modern ICs that uses islands of clock synchronous logic (e.g. individual IP cores) that are connected asynchronously (i.e. avoiding a global clock) [63].

guaranteed latency (GL) Guranteed latency denotes the property to require an upper bound/limit on the latency or response time. An example is network traffic where the network packets must arrive at the destination before a certain deadline.

guaranteed throughput (GT) Guranteed throughput denotes the property to require a lower bound/limit on the accepted throughput. An example is network traffic where the network packets must arrive at the destination with a certain rate.

latency The delay experienced by an entitiy (e.g. frame, packet, or flit) in the course of its propagation between two points in a network, measured from the time that a known reference point in the frame passes the first point to the time that the reference point in the frame passes the second point [11].

maintainability The ability to be retained in, or restored to a state to perform as required, under given conditions of use and maintenance Given conditions would include aspects that affect maintainability, such as: location for maintenance, accessibility, maintenance procedures and maintenance resources [52].

mixed-criticality Function of different safety level, e.g., safety-critical and non safety-critical in the same system. In this context, a critical function is essential for the safety of the system. Therefore, this function is developed with high diligence and so the behaviour (i.e. timing) is well specified and tested. For non-critical functions the confidence in the characteristics is lower, i.e., the possibility that the function deviates from the specification is higher.

multiprocessor system on chip (MPSoC) A system-on-chip (SoC) that includes multiple microprocessors/processing units.

network interface (NI) The network interface connects a node to the NoC.

network interface unit (NIU) see NI.

network-on-chip (NoC) A network-on-chip (or on-chip network) is a network-based communication subsystem in an integrated circuit, such as, for example, in a system-on-chip (SoC).

predictability The capability to dervie a consistent repetition of a state, course of action, behavior, or the like, making it possible to know in advance what to expect. Predictability is the degree to which a correct prediction or forecast of a component's or system's state can be made either qualitatively or quantitatively. Formal methods (e.g. analysis) can be used to predtict the bahaviour of a system or component, see Analysability.

quality of service (QoS) The collective effect of service performances, which determine the degree of satisfaction of a user of the service. These characteristic performances may, for example, relate to: transmission quality, timing, failures, fault frequency and duration [52].

real-time system A system which is able to manage tasks continually, so that it is possible to react as directed on process events in a predetermined time period [52]. For a real-time (or time-critical) application or syste the correctness of the application or system function depends not only on functional but also on temporal aspects.

recoverability The ability to recover from a failure, without corrective maintenance. The ability to recover may or may not require external actions. For recovery where no external actions are required, see self-recoverability. Recoverability may be quantified using measures such as the probability of recovery within a specified time interval [52].

reliability The ability to perform as required, without failure, for a given time interval, under given conditions. The time interval duration can be expressed in units appropriate to the item concerned, e.g. calendar time, operating cycles, distance run, etc., and the units should always be clearly stated. Given conditions include aspects that affect reliability, such as: mode of operation, stress levels, environmental conditions, and maintenance [52].

safety Freedom from unacceptable risk of physical injury or of damage to the health of people, either directly, or indirectly as a result of damage to property or to the environment [15].

self-recoverability The ability to recover from a failure, without external action. Self-recoverability is a special case of recoverability [52].

stream A unidirectional flow of data (e.g., audio and/or video) from a sending network node (*Talker*) to one or more network nodes (*Listeners*) [11].

sufficient independence Sufficient independence of implementation in a mixed-criticality system is established by proving that timing interference or the probability of a dependent failure between the non-safety and safety-related parts is sufficiently low in comparison with the highest safety integrity level associated with the safety functions [15]. While a failure can result from, for example, a fault, wilful timing attack, or wilful memory manipulation and influence timing, data consistency, or other parameters of the system.

system-on-chip (SoC) An integrated circuit, which contains all the required components of an electronic system on a single chip [50].

virtual channel (VC) A virtual channel is a separate FIFO buffer inside the input module of a network switch [54; 63; 69].

worst-case execution time (WCET) Maximum possible time during which a program is actually executing [8]. In scope of this thesis, the worst-case execution time of a flit at a network router describes the maximum possible time the router is actively procesing the flit exluding blocking through other flits.

worst-case responce time (WCRT) Maximum possible time between receiving a stimulus and delivering an appropriate response or reaction [8]. In scope of this thesis, the worst-case response time of a flit at a network router describes the maximum possible time between the arrival of the flit at the router and its forwarding including blocking through other flits.

Acronyms

ADAS advanced driver assistance system.

BCET best-case execution time.
BCRT best-case response time.
BE best-effort.

COTS commercial off-the-shelf.
CPA compositional performance analysis.
CPU central processing unit.
CT control traffic.

DSP digital signal processor.
DTS distributed traffic shaping.

ECU electronic control unit.

flit flow control digit.

GALS worst-case response time.
GL guaranteed latency.
GPU graphics processing unit.
GT guaranteed throughput.

MPSoC multiprocessor system on chip.

NI network interface.

NIU network interface unit.
NoC network-on-chip.
NoC-RM network-on-chip resource management.

QoS quality of service.

RTE runtime environment.

SoC system-on-chip.

VC virtual channel.

WCET worst-case execution time.
WCRT worst-case response time.

www.ingramcontent.com/pod-product-compliance
Lightning Source LLC
Chambersburg PA
CBHW060249220326
41598CB00027B/4032